More
Telephone
Accessories
You Can Build

More Telephone Accessories You Can Build

JULES H. GILDER

HAYDEN BOOK COMPANY, INC.
Rochelle Park, New Jersey

Patent Notice

Purchasers and other users of this book are advised that projects described herein may be proprietary devices covered by patents owned or applied for and that their inclusion in this book does not, by implication or otherwise, grant any license under such patents or patent rights for commercial use. No one participating in the preparation or publication of this book assumes responsibility for any liability resulting from unlicensed use of information contained herein. Information furnished by the author is believed to be accurate and reliable. However, no responsibility is assumed by the author or publisher for its use.

Library of Congress Cataloging in Publication Data

Gilder, Jules H

   More telephone accessories you can build.

   1. Telephone—Apparatus and supplies—Design and construction—Amateurs' manuals.   I. Title.
TK9951.G53   621.386   80-24943
ISBN 0-8104-0893-7

Printed in the United States of America

2	3	4	5	6	7	8	9	PRINTING
81	82	83	84	85	86	87	88	YEAR

Preface

The one common household appliance that holds more fascination for the average person than any other is the telephone. With this small device it is possible to talk to people all over the world from the comfort of your own home. People have always been interested in expanding the use of the telephone in their day-to-day lives, as witnessed by the explosive growth of the telephone accessories market. But most of these accessories tend to be expensive, and quite often, they don't do exactly what you'd like them to.

This book is designed to help you overcome some of these problems by providing a variety of telephone projects that you can build inexpensively. While some are home-built versions of commercially available products, others—such as the Busy Phone, Call Length Limiter, and Priority Telephone, to name a few—are not yet available on the commercial market.

Unlike most of the projects in my earlier book, *Telephone Accessories You Can Build,* a large number of these projects require a direct connection to the telephone line. The telephone company usually frowns on this practice and frequently requires that you install an adapter between your device and their phone line. This is a safeguard meant to protect their equipment and personnel. Therefore, you are advised to check with your local telephone company to determine if such a device is necessary before any of the projects in this book are connected to the telephone line.

Since telephone networks throughout the country are constantly changing, the characteristics of the lines going to the individual subscriber's phone occasionally change as well. While all of these projects have been built and tested in advance, it is possible that an occasional one will not work in your local area because of the peculiarities of your local phone system. If you have any suggestions for circuit improvements or new projects, I would be happy to hear about them.

One final point, it is suggested that you power all your circuits with batteries rather than a dc power supply in order to eliminate the possibility of any ac voltage setting onto the line and endangering telephone company employees and equipment.

JULES H. GILDER

Contents

More Telephone Accessories You Can Build

1

Conference Caller

Would you like to be able to talk to your Aunt Bessie in Florida and your sister Eileen in New York at the same time? You could call the operator and, for a special fee, probably arrange such a round robin, commonly referred to as a *conference call*. Or, if you're lucky enough to live in an area where the phone company is offering a special conference capability to its subscribers for an additional monthly fee, you could set up such a call yourself.

If neither one of these solutions to the problem appears to be practical, then rummage through your junk box for a handful of parts or go down to your nearest electronic parts store and buy everything you need to make a conference calling device for less than $5 — nothing more than two 1-μF nonpolar capacitors and a double-pole-single-throw (dpst) switch. If you want your project to look nice and simple to connect up, you can also buy a 3 × 2 × 1-inch plastic box and a telephone jack /plug.

One final important thing: To use this device, you must have two separate phone lines (two different numbers), and one of them should have Touch-Tone™ capability. Also, since the device will be directly connected to the phone line, you may wish to check with your local telephone company to see if they have any objection.

About the Circuit

Talk about simple circuits — you can't get much simpler than this one. All it consists of is two capacitors and a switch. Even a neophyte electronics hobbyist should have no problem putting this handy telephone accessory together.

The basic theory behind the circuit is to couple two telephone lines without interfering with the operation of either one. What we really want to do is let the audio signals pass from one line to the other while preventing the dc control signal of the telephone network from passing. One way to do this is with a specially wound isolation transformer whose windings are of equal impedance. A simpler and cheaper way is simply to put a 1-μF capacitor in series with the connections between the two telephones, as illustrated in Fig. 1.1. The switch makes it possible for you to connect the second line or cut it off whenever you want to.

1

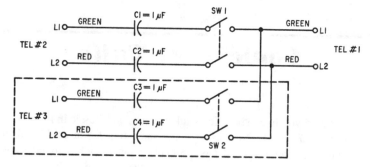

Fig. 1.1   Conference caller

Construction and Installation

Construction of this device is simple and wiring is not at all critical. Components can be mounted directly to the switch, eliminating the need for a circuit board. When connecting the conference caller to your telephone lines, you must pay attention to the polarity of the wires. Make sure that only the green wires get connected to the green terminals and the red wires to the red terminals.

Another thing to make sure of is that capacitors C1 and C2 are nonpolar (not electrolytic) capacitors. They should also have a working voltage of at least 100 V. If you wish to expand your conference caller to work with three or more lines, simply add two more capacitors and a switch for each additional line (dashed box in Fig. 1.1.).

To test out the conference caller, have someone call one of the numbers connected to the device. Next, call someone else on the other phone that is connected to the conference caller. Now close the switch and all three of you should be able to talk together. Happy conferencing and welcome to the world of improved telephone communications.

Parts List

C1–C4	1-μF nonpolar capacitor (100 WVDC)
SW1–SW2	dpst switch

2

Music on Hold

Have you ever called a doctor's office and been told to hang on for a minute? If so, chances are that while you were waiting you suddenly heard some background music meant to entertain you and help keep your attention from the dragging time. Now, with this button, you too can place people calling you on hold and even let them listen to music until you get back to them. You will be able to answer the phone in one room of your house, put the caller on hold, and then pick up the receiver at another location. When you pick up the receiver the second time, you automatically deactivate the music-on-hold feature and can continue your conversation undisturbed. All these advantages can be had for less than $10.

This project requires a direct connection to the telephone line. Note that some telephone companies object to customers making such connections, fearing that they might introduce high voltages that could be harmful to telephone company employees or equipment. This device does not use such voltages. If in doubt about your telephone company's position, however, check with them before making any direct connections to the line.

About the Circuit

This circuit (Fig. 2.1) is a relatively simple one and ideal for the not-so-experienced hobbyist. It has a total of seven electronic components and can be put together in less than an hour by even a novice.

The basic operation of a hold button requires that a high resistance — about 1200 ohms — be placed across the telephone line while it is in use to prevent the line from being disconnected when you hang up the receiver. (In some parts of the country, lower resistance values may be needed.)

Now, if part of this high resistance is formed by the secondary of an inversely connected transistor output transformer, then an audio signal can be coupled into the telephone line for the benefit of the party on hold.

In operation, this seven-component circuit is connected in parallel with the telephone line. When the receiver is lifted off the hook, the voltage on the line is about 5 V. Even if the hold button were activated, this voltage would be too low to keep it so. If the pushbutton were pressed and the receiver hung up, however, the voltage on the line would rise to

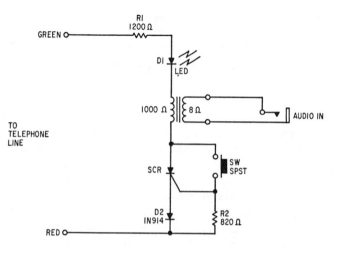

Fig. 2.1   Music on-hold circuit

about 48 V. R1, T1, and D1 momentarily form a voltage divider with R2, allowing part of the line voltage to be applied to the gate of the SCR and triggering it into its conducting mode.

The triggered SCR acts as a short circuit that connects the resistor, LED, transformer, SCR, and diode series circuit across the phone line, thus raising the resistance across the line to between 1200 and 1500 ohms and placing the line on hold. In addition, if an audio signal, such as that obtained from a radio or tape recorder, is fed into the 8-ohm primary of the transformer, the signal will be coupled into the telephone line and the person waiting on hold will hear it.

When the telephone or any extension is subsequently picked up, the line voltage drops again to about 5 V and the SCR becomes current-starved, thereby causing the SCR to stop conducting and effectively opening the circuit and disconnecting the phone line from the hold mode and the audio source.

Construction

As mentioned earlier, the small number of components required for this circuit makes construction quick and easy. The whole unit can be built into a 3 × 2 × 1-inch plastic box. Because of the unit's simplicity, no printed circuit board is needed. In fact, if desired, the transformer can be glued to the lid of the box and the remaining components mounted via their connections to the switch or the LED.

There are a few simple but important points to which you must pay attention. The first is the polarity of the LED. The LED must be connected so that the anode of the device goes to the positive (green) wire of the telephone; the cathode, to the red wire. The next thing to look out for is the polarity of diode D2. Its cathode must go to the red wire of the telephone, along with one side of the 820-ohm resistor.

Since component values are not critical, just about any silicon diode can be used for D2, as well as for the SCR and the LED. Resistor R1 may need some adjustment to compensate for different values of SCR conduction current.

Installation and Operation

There are two ways in which the music-on-hold button can be connected to your telephone. If you wish, the unit can be wired directly to your wall jack. A more convenient, although slightly more expensive approach, is to use a jack/plug. The latter will make it possible to disconnect the unit quickly whenever you wish and to move it easily from one location to another (making sure that the red and green wires remain properly connected).

To test out the unit, have a friend call you and tell him that you are going to put him on hold for a minute but that you will be right back. Next, press the hold button and hang up the phone while the button is down. If you have previously connected an audio signal to the transformer, the moment you press the button you should hear the audio in the receiver of your telephone. Otherwise, the audio has not been successfully coupled into the telephone line.

The music will continue to be heard by the person on hold until you once again pick up the receiver. As a reminder to you, the LED glows brightly all the time that someone is on hold, extinguishing only when you pick up the phone again.

Parts List

D1	light-emitting diode (see text)
D2	1N914, or any silicon diode
R1	1200-ohm resistor (see text)
R2	820-ohm resistor
T1	1000- to 8-ohm transformer
SCR	RS 1020 (Radio Shack)
SW	spst momentary-contact pushbutton

3

Music Synthesizer on Hold

For those of you who were interested in the music-on-hold circuit, this hold circuit should strike your fancy too. The former circuit permitted you to connect a radio, tape, or other source of audio to your phone line for the benefit of the calling party on hold. For those of you with a musical bent or who just don't want to tie up a radio or tape recorder, we will now consider a musical synthesizer that will play a 64-note tune continuously until a party is taken off hold.

Theory of Operation

The basic requirements for placing an incoming call on hold were described in the previous section and will only be summarized here. To place a party on hold, you've got to simulate an off-hook situation by placing a resistance of 300 to 1200 ohms across the telephone line. A convenient value to use is something in between, say 600 ohms, since it is fairly easy to find 600-ohm relays.

Unlike the previous circuit, which was all solid-state and used an SCR as the latching element, this circuit uses a two-pole relay. One set of contacts of this relay is used to latch the relay closed after the hold button is pushed, and the other set is used to turn on the music synthesizer.

The synthesizer circuit (Fig. 3.1) can be divided into three stages. The first is the astable multivibrator that is fabricated from an IC timer circuit — the 555. The oscillator frequency can be adjusted by potentiometer R2 and operates in a range of 1 to 10 Hz.

The second stage feeds the output of the astable multivibrator to a divide-by-64 counter that is fabricated from three dual J-K flip-flops. The normal (Q) outputs of these flip-flops are connected to the toggle input of the following stage, while the inverted outputs ($\bar{Q}$) are buffered and summed at the base of transistor Q1. Before they are summed, however, these signals are passed through a variable resistor so that the amplitude of the pulses can be changed. The resulting signal at the base of $\bar{Q}$1 is a sum of all the currents. (Note that although the unit described is designed for 64 notes, more or less notes can be obtained by increasing or decreasing the number of flip-flop outputs that are summed.)

6

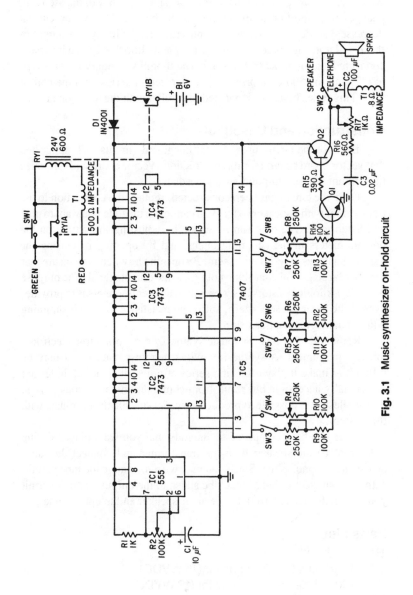

Fig. 3.1 Music synthesizer on-hold circuit

The last stage of the synthesizer circuit is composed of Q1 and Q2, which form a current-controlled oscillator. The frequency of the oscillator changes in proportion to the current at the base of Q1. As the current decreases, so does the frequency, and vice versa. Finally, the output of this oscillator stage is coupled to the telephone line through an inversely connected audio-output transformer. The 8-ohm secondary is connected to Q2, and the 500-ohm primary is connected in series with the coil of RY1. This series circuit is connected in parallel with the telephone line.

Construction and Checkout

Because the circuit uses several integrated circuits, it is suggested that vector board be used for its construction. Vector board will minimize the amount of point-to-point wiring needed.

Once the circuit has been constructed, replace the connection to the transformer with a speaker so that you can set up the tune you want to play and listen to it. The tune can be varied by adjusting the setting on the potentiometers in the summing circuit and by opening or closing the various switches in series with them. By opening a switch, you insure that a particular stage will contribute nothing to the output signal. The purpose of the 100-kilohm resistors in series with the potentiometers is to provide, even if the latter are set to zero, sufficient resistance to permit summing without damaging the transistor.

Remember that this is a music synthesizer and not a true electronic musical instrument. A synthesizer will generate music, but it may be difficult to make it play a specific series of notes since not all notes are individually accessible but result from combinations of various pulses. Nonetheless, it is possible to produce very pleasing electronic music with this system.

After you have composed the melody that you want played during hold, replace the speaker with the transformer and connect the latter across the telephone line. Since you may wish to change the tune periodically, it is suggested that a switch be added to the output that will permit you to switch back and forth between the speaker and telephone line.

Parts List

B1   6-volt battery
C1   10-μF electrolytic capacitor (16 WVDC)
C2   100-μF electrolytic capacitor (16 WVDC)
C3   0.02-μF disc capacitor
D1   IN4001 silicon diode
IC1 555 IC timer

IC2–IC4	7473 dual J-K level-triggered flip-flops
IC5	7404 hex inverter
Q1	2N3904
Q2	2N3906
R1	1-kilohm, ½-watt resistor
R2	100-kilohm potentiometer
R3–R8	250-kilohm potentiometers or trimpots
R9–R14	100-kilohm, ½-watt resistors
RY1	24-volt, 600-ohm dpst relay
SPKR	8-ohm speaker
SW1	momentary push button switch (spst)
SW2	toggle or slide switch (spdt)
SW3–SW8	toggle or slide switch (spst)
T1	audio output transformer (500-ohm impedance primary, 8-ohm impedance secondary)

4

Visual Ring Indicator

If you are trying to sleep and the telephone rings, it can' be very annoying. You can, of course, prevent the phone from ringing by taking the receiver off the hook. However, since the telephone company does not like you to leave the receiver off the hook because you may forget it in that position, after a few minutes they generate a loud audio signal to try to catch your attention. Not exactly conducive to a nice comfortable sleep. Besides, other members of your family may be expecting a phone call and not appreciate your ploy.

How do you get around this problem? Simple. Just build this visual ring indicator and silencer. When this device is connected to your telephone, a flick of the switch will disconnect the ringer and connect the LED instead. Now any time the phone rings, the bell will be silent but the LED will light up.

About the Circuit

If you take a look at the way your telephone is connected to the jack or junction box in your home, you will notice that the green and yellow wires are usually connected together, whereas the red one is separate. When the telephone company wants to ring your bell (except on party line installations), they send a 90-volt, 20-Hz signal down your line between the green and red wires. When it gets to your telephone jack, this signal is sent to the bell mechanism via the yellow wire and causes the phone to ring.

In this project, we simply break the direct connection of the yellow wire to the green one by a single-pole, double-throw switch (Fig. 4.1). In one position, which is the normal case, the connections to the switch are arranged so that the yellow wire is shorted to the green, thus allowing the telephone to ring. In the second position, the connection to the yellow wire is broken and the green wire is connected instead to a series circuit consisting of a 10-kilohm resistor, a silicon diode, and a light-emitting diode.

Since about 48 volts are normally across the line when the receiver is on the hook, you might expect the LED to remain lit all the time. This would be the case if the LED were connected with a reversed polarity to

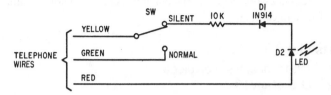

Fig. 4.1.   Visual ring indicator

the line. Because of the way it is connected in Fig. 4.1, however, the LED is normally reverse biased and thus is not lit.

Since the ring signal that activates the bell is an ac signal, on positive half cycles it will not do anything, but on negative half cycles, the voltage will cause the LED to be forward biased and thus light up. The LED will thus flash on and off with each ring.

Construction and Installation

If you so desire, the circuit is simple enough for you to build right into your telephone. If this is the route you choose, make sure that the telephone belongs to you and not to the telephone company. Phone companies frown on customers making physical modifications to their telephone instruments. Desk phones are readily available in stores for $8 to $10 and for as low as $3 in flea markets.

To build the visual ring indicator into a telephone, you will need to drill two small holes to accommodate the switch and the LED. The other two components can be wired directly to these devices and be supported by them.

If you are going to build the indicator in a separate little box, you will need to have about 3 feet of two-conductor wire to go between the box and the phone jack. And if you wish, you can connect this to your phone lines with a simple plug and jack. Wiring is fairly simple, and the whole job should take you no more than half an hour to complete.

One thing that you should pay particular attention to is the polarity of the LED and the silicon diode. If the polarity of either one of them is reversed, the LED will not light up at all, and if the polarity of both of them is reversed, the LED will remain on all the time.

To test out the unit, simply have someone call you and put the switch in the silent position *after* the first ring. You should leave it in the normal position until the first ring is completed so that, if for some reason the device does not work, you'll know that it is its fault and not the failure of your friend to call you.

Parts List

D1 1N914, or any silicon diode
D2 LED of any kind
 R 10,000-ohm, ½-watt resistor
SW Spdt switch

5

*Phone Line Monitor**

For the phone extension that has no bell, this "phone-line" monitor watches the flashing buttons of a two-line phone extension and provides an audible tone.

A 2N5777 photodarlington cell picks up a blinking light from the transparent plastic buttons. The cell mounts to the phone with a black adhesive tape (Fig. 5.1) and is connected to a tiny circuit board by a flexible cord that can be taken from a transistor-radio earphone.

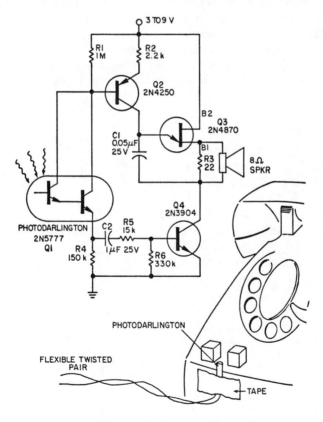

Fig. 5.1. Phone line monitor

Special features of the monitor circuit include low cost, no electrical connection to the phone line, and current consumption only when the light flashes. The power is switched on and off by a hi-beta 2N3904 transistor. The circuit's 9-V battery can be left continuously connected Less than a microampere of current is drawn, even with normal ambient light and the phone light not flashing.

The logarithmic action of a 2N4250 transistor produces easily detectable differences in the pitch and pattern of output tones. The differences allow you to know whether two lines are ringing, one is on hold and the other ringing, and so on. Differences result because the hold-light pattern is quicker than the ringing pattern.

Because of the circuit's coupling, the output tones are soft and gentle but stirring enough to command notice. When a line is called, the speaker emits an initial high-pitched signal that subsides quickly to a lower, but still noticeable, pitch. On hold, the tone continues to pulsate, thus serving as a good reminder for a secretary, say, that someone is still waiting on the line. The tone subsides completely when a line is answered or used for calling, since the light then remains on continuously.

For noisy locations, the tone can be made louder with an input transformer (ratio of 250:8) or a 100-ohm speaker that replaces the 22-ohm resistor in the output.

Parts List

C1	0.05-μF, 25-V capacitor
C2	1-μF, 25-V capacitor
Q1	2N5777 photodarlington transistor
Q2	2N4250 pnp transistor
Q3	2N4870 unijunction transistor
Q4	2N3904 npn transistor
R1	1-megohm resistor
R2	2.2-kilohm resistor
R3	22-ohm resistor (see text)
R4	150-kilohm resistor
R5	15-kilohm resistor
R6	330-kilohm resistor
SPKR	8-ohm speaker (see text)

*H. MacDonald, "Circuit Monitors Blinking Phone Lights and Provides Soft But Commanding Tone," *Electronic Design* 22, October 25, 1976, p. 194.

6

Missed Call Indicator

Sometimes while waiting for an important call to come in, you may have to leave the telephone for a little while. When you return, you have no way of knowing if anyone called. With this handy little circuit, however, you do. It will monitor your telephone and let you know if someone called. Of course, if you had a telephone answering service or machine, you'd also be able to know who called. But this unit may also be handy even if you have an answering machine because many have no way of telling you if someone called unless you listen to them to find out. This device, however, will turn on a light that tells you right away that someone has called.

Theory of Operation

The operation of this circuit (Fig. 6.1) is really quite simple. When its number is called, an electromagnet inside the telephone is energized and operates a clapper that rings the bell. Now, if a magnetically sensitive switch were to be located next to this electromagnet, then it too could be activated each time the telephone rings. Such a switch is known as a *magnetic reed switch*. Frequently used in low power relays as well as burglar alarms, it is available in most electronic supply houses.

The reed switch is hooked up so that when it closes, it applies a voltage to the gate of the SCR, turning it on. Once on, the SCR light up the LED, which stays on until either SW1 is pushed, resetting the circuit, or SW2 is turned off.

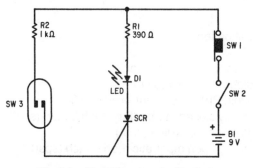

Fig. 6.1. Circuit for missed call indicator

Putting It Together

Construction of this device is not at all critical; just about any SCR, LED, and reed switch will work nicely. The parts (except for the reed switch) can either be mounted on a piece of perforated board or soldered directly to the switches, which can be mounted in a small box. The reed switch is attached to two pieces of wire about 15 inches long. After the connection is made, it must be insulated with electrical tape so that none of the metal that is part of the connection is exposed. The other ends of the two wires are connected to the free end of R2 and the gate of the SCR, respectively. To make things a little more convenient, you may wish to use a miniature plug and jack here.

Checking It Out

Once the unit is constructed, you can do a very simple preliminary check by turning it on and bringing a magnet next to the reed switch. By doing this, you should cause the switch to close, and the LED should light and stay lit until you press SW1. If the LED does not light, check the polarity to make sure that it is connected correctly. Once the unit checks out properly, open your telephone by loosening the two screws on the bottom. It is best to do this on a telephone that you have purchased privately, because telephone companies frown on anyone tampering with their equipment. Once the screws have been loosened, remove the case and look for the bell ringing mechanism, part of which is an electromagnet. Place the reed switch next to this electromagnet, and close SW2 on the indicator circuit. The LED should not light. If it does, there is some residual magnetism in the electromagnet core, and you'll have to move the reed switch a little further away from the electromagnet. Once you have located the switch, have someone call you and make sure that the LED lights up after the first ring. If it doesn't adjust the position of the reed switch until it does.

Parts List

B1	9-volt battery
D1	LED
R1	390-ohm, ½-watt resistor
R2	1-kilohm, ½-watt resistor
SCR	25-volt, PIV silicon controlled rectifier
SW1	normally closed momentary push button (spst)
SW2	toggle or slide switch (spst)
SW3	normally open magnetic reed switch (spst)

7

Call Length Limiter

If you've got a teenager in the house and only one telephone, chances are that you won't get to use the phone very often because it is always tied up. Now, through the magic of electronics, you can limit anyone's calls on the telephone to ten minutes (or any time period you wish).

The call length limiter is a simple circuit that is connected in parallel with the telephone line and operates with all extensions. Where it is activated — a key switch to turn it on and off is suggested — a person using the telephone has only ten minutes to complete his call, after which a steady tone is introduced onto the line, making further conversation almost impossible. The tone doesn't come without warning. After eight minutes, a short warning beep announces that only two minutes are left.

The limiter isn't restricted to outgoing calls only, moreover; it will work just as well on incoming calls. Once the tone comes on, it can be stopped only by turning off the key switch. With a slight modification of the circuit, the tone can be caused to shut off by hanging up the receiver. On outgoing calls the tone forces the conversation to come to an end, but on incoming calls, it does not necessarily terminate them, because it is possible for the called party to hang up the phone for a short while and then pick it up again without terminating the connection.

Theory of Operation

The call length limiter circuit consists of a unijunction transistor and two 555 integrated circuits. The unijunction transistor is a three-terminal device that, as its name implies, has only one pn junction. Under certain conditions, the unijunction transistor can exhibit a negative input resistance — a characteristic that makes it usable in timing applications.

In the basic unijunction oscillator circuit (Fig. 7.1), the B1-E junction is reversed biased and no current flows through it at the very beginning. The timing capacitor charges up through R1. When the voltage on the capacitor reaches a threshold value, the emitter junction becomes forward biased and current starts to flow through it to B1, thereby reducing the internal resistance of the unijunction transistor. The charge now stored on the capacitor is dumped into the load resistor, R1,

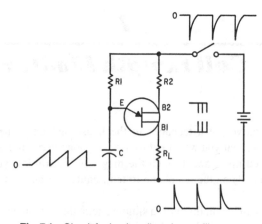

Fig. 7.1   Circuit for basic unijunction oscillator

thus producing three usable signals: a positive-going spike at B1, a negative-going spike at B2, and a sawtooth waveform at E.

In the call length limiter, a unijunction device is used to generate a positive-going pulse after a delay of eight minutes. Here's how it works. As long as the telephone receiver is on the hook, 48 volts is across the line. The diode bridge used at the input to the circuit makes the circuit impossible to connect incorrectly on a two-wire system. Either lead (A or B) can be connected to the green wire from the telephone, with the other one going to the red wire. In either case, −48 volts appears at the cathode of the 9-volt zener diode (D5). Since the −48 volts is greater than the 9-volt barrier presented by the zener diode, the diode conducts and the −48 volts is applied to the junction of R1 and C2. A negative voltage at this junction keeps unijunction transistor Q1 cut off.

When the telephone handset is lifted off its cradle for either an incoming or outgoing call, the voltage on the telephone line drops to about 6 volts. When this happens, the zener diode (Fig. 7.2) stops conducting, and C2 starts to charge up to a positive voltage via resistor R1. When the threshold voltage is reached, after about eight minutes, the charge on the capacitor is dumped to B1, creating a positive-going pulse. This pulse is sent to two locations. It is sent via isolation diode D7 to IC2, which is configured to work as a 1-kHz oscillator. The output of IC2 is connected to a reverse-connected audio output transformer, whose secondary is coupled to the telephone line. The short pulse causes the oscillator to produce a brief beep on the telephone line, warning the user that only two minutes remain before a steady interrupting tone will occur.

The pulse from B1 is also coupled to an SCR. The SCR is connected between the positive power supply and IC1. When a positive pulse

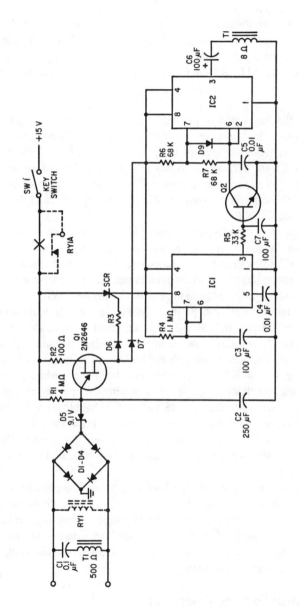

Fig. 7.2. Circuit for call length limiter

appears on its gate, the SCR turns on and locks on, supplying power to both IC1 and IC2.

IC1 is configured as a special type of monostable multivibrator called a *power-up monostable*. When power is applied to this circuit, the output on pin 3, which until now has been low, goes high for a period of time determined by R4 and C3. The time is determined by the equation

$$t = 1.1 \,(R4)(C3)$$

where the resistance is in ohms; the capacitance, in farads; and t, in seconds. In this case, the period of time is about two minutes. The positive output signal of IC1 is applied to the base of npn transistor Q2 via R5 and C7, which causes a 3-second delay before turning the transistor on and causing it to short C5, the timing capacitor of IC2. As long as this capacitor is shorted, IC2 does not operate, and thus no tone is produced.

After two minutes, the output of IC1 returns to its normal low state, transistor Q2 is cut off, and capacitor C5 starts to charge normally. This allows IC2 to oscillate and produces a steady 1-kHz tone that is coupled to the telephone line by the audio transformer. IC2 will continue to oscillate until power is removed from the circuit by opening SW1.

A modification can be made to the circuit that will simplify its operation. Shown if Fig. 7.2, it is designed to reset the call limiter automatically whenever the telephone handset is hung up, thereby eliminating the need to turn off the limiter and restart it time and again with the key switch. As can be seen from the dashed lines in the schematic, the change is simple. It merely involves putting a 48-volt relay in parallel across the line. If a 48-volt relay is hard to come by, just use a 24-volt relay in series with a resistor that has the same value as the coil resistance of the relay. The relay should have a set of normally closed contacts. These contacts, as can be seen, are connected in series with the key switch.

When the phone is on the hook, no power is applied to the limiter circuit. When it is off the hook, however, the relay drops out and the contacts return to their normally closed position, powering up the limiter circuit. When the 1-kHz tone comes on the line, it may be stopped simply by hanging up the phone or by opening the key switch. Both actions interrupt power to the circuit.

Construction and Checkout

As is true of most projects in this book, the circuit is relatively simple and can be constructed on a piece of breadboard. There are quite a few semiconductor parts, and it is particularly important to pay attention to the polarity of the various devices, especially the diodes. Even one incorrectly inserted diode can make the entire circuit fail.

If the circuit does not work the first time out, first check to see if there are −48 volts at point C in the schematic. If not, you probably hooked up the bridge circuit incorrectly or the phone is off the hook. If there are, check to see if there are −48 volts on capacitor C2. If not, the zener may be rated too high. If there are, lift the receiver off the hook and monitor the voltage of C2 with a high input impedance voltmeter or scope. It should be constantly increasing. Or wait eight minutes and see if a pulse is generated on B1. If it is, check the voltage on the cathode of the SCR. It should be close to 15 volts. If there is no voltage there, either no pulse was generated or the SCR is not operating properly. With a voltage at the cathode of the SCR, IC1 should be working. Check this out by measuring the voltage across C5. It should be zero for two minutes. After that, it should start oscillating.

To test the unit in operation, simply connect it to the telephone line. Don't worry about which wire is connected where because the diode bridge at the input to the circuit makes it insensitive to polarity. Next, lift the receiver off the hook and call someone. Have a watch handy so that you can check the time. You should hear a beep after eight minutes and a steady tone after another two.

If you want to make these time periods shorter or longer, all you have to do is change either capacitor C2 or C3. Increasing C2 gives you a longer initial period before the warning beep is generated. Increasing C3 give's you more time between the warning beep and the steady tone. The time delay in both cases is linearly proportional to the capacitor values.

If you've got a really talkative person in the house, the circuit should pay for itself in one month.

Parts List

C1	0.1-μF capacitor
C2	250-μF electrolytic capacitor (25WVDC)
C3	100-μF electrolytic capacitor (25WVDC)
C4	0.01-μF capacitor
C5	0.01-μF capacitor
C6, C7	100-μF electrolytic capacitor (25WVDC)
D1–D4	silicon diode bridge (100 PIV)
D5	9.1-volt zener diode
D6–D9	1N914 silicon diodes
IC1, IC2	555 IC timer
Q1	2N2646 unijunction transistor
Q2	2N3904 general-purpose npn transistor

R1	4-megohm, ½-watt resistor
R2–R4	100-ohm, ½-watt resistors
R5	32-kilohm, ½-watt resistor
R6–R7	68-kilohm, ½-watt resistors
RY1	48-volt, 1200-ohm relay
SCR	silicon-controlled rectifier (1 amp, 25 PIV)
SW1	key switch (spst)
T	8- to 500-ohm audio output transformer

8

Busy Phone

Did you ever go into the shower and all of a sudden hear the phone ring? Or did you ever sit down to a nice relaxing meal when all of a sudden the phone rings? How many of us have the self control to let it continue to ring until we finish? Was there ever a time when you wished you didn't have to worry about the telephone interrupting you but at the same time didn't want people to think you weren't home? Most of us can answer yes to at least one of these questions. And if you can, then the Busy Phone is for you.

Some people will say that all you have to do is take the telephone off the hook, and they are right. But it is not such a simple matter to remember to put it back on again later. Many times people leave the phone off the hook for an entire day without realizing it, having forgotten to put it back on when the period in which they did not want to be disturbed ended.

The Busy Phone takes all the remembering out of taking the phone off the hook. All you have to do is press a button, and the Busy Phone simulates an off-hook condition. Anyone trying to call you will get a busy signal, and after a maximum of one-half hour, the circuit will automatically disarm itself and return your telephone to its normal on-hook condition.

The Busy Phone has two more features that make it convenient to use. The first is an indicator that it is in operation and working properly, and the second is a reset button that will permit you to return the phone to its normal condition even before the half-hour busy period has elapsed.

Theory of Operation

Like so many other projects in this book, Busy Phone uses the extremely versatile 555 intergrated circuit timer. The timer IC is connected up as a monostable multivibrator (Fig. 8.1). The specific configuration used is a power-up monostable. It operates in the following manner. When switch SW1 is pressed, it applies power to the IC. C1 is initially discharged, making the trigger input to the IC (pin 2) active. The trigger input causes the output, at pin 3, to go high, thereby supplying power to the relay. The relay closes and, through contacts RY1A, locks power onto

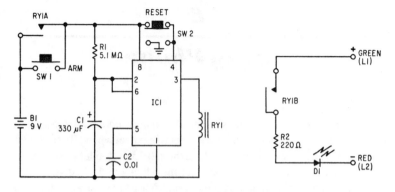

Fig. 8.1 Circuit for Busy Phone

the 555 timer IC, even though the momentary push button SWI has returned to its normally open position. With the power applied, capacitor C1 charges up through resistor R1. When the voltage on the capacitor reaches 6 volts, the upper comparator inside the 555 (see Appendix) resets the internal flip-flop and drives the output, pin 3, low, thereby turning off relay RYI. The 6-volt level is reached after a time period determined by the equation

$$t = 1.1 (R1)(C1)$$

With the components shown in the diagram, the maximum time delay produced by the circuit is a little more than 30 minutes.

It is possible that the time delay will be longer than anticipated if cheap electrolytic capacitors are used, the reason being that such capacitors have charge leakage that makes them take longer to charge than anticipated. In any case, as long as the voltage across the capacitor is less than 6 volts, the relay will be on.

If you wish to terminate the timing cycle before it is completed, it is necessary only to push SW2. This switch is connected to the reset terminal of the timer IC. Under normal conditions, this terminal is inactivated by tying it up to the positive voltage supply. However, if it is brought down to zero potential, as it is when SW2 is pushed, the output flip-flop inside the 555 is reset, thus forcing the output, pin 3, to go low no matter what state the other inputs to the flip-flop are in.

Until now we have merely discussed the operation of the monostable multivibrator and not what happens to the telephone line. Explaining that is simple. When the relay closes contacts RY1A, it also closes another set of contacts, RY1B. This set of relay contacts is connected across the telephone line in series with a resistor, R2, and a light-emitting diode (LED). The resistor does two things. First, it limits the amount of current that can flow through the LED so that it will not be damaged. And second, and more important, it puts a low resistance across the telephone line, causing the telephone central office equipment to think that the telephone is off the hook. Thus, whenever the Busy Phone is in use, the LED is on to indicate this fact, and a low resistance is placed across your telephone line so that anyone who is calling will get a busy signal.

Construction and Checkout

Construction of the Busy Phone is really quite simple. It may be built on a piece of phenolic or vector board and should be housed in a small minibox, with the two switches and LED mounted on the front panel and a dual conductor coming out the side for connection to the telephone network. For ease of use, a modular or jack connector should be used.

While the circuit is simple, there are two points to which some attention must be paid. The first is capacitor C1. In addition to taking the normal precaution of making sure that it is hooked up with the proper polarity, you should try to use a low-leakage capacitor, such as a tantalum one. This will insure more accurate time delays.

The other thing to which some attention should be paid is the way the circuit is hooked up to the telephone line, since it is polarity sensitive. The cathode of the LED must go to the negative side of the line, which is generally the red wire. If you are not sure, take a voltmeter and connect it across the telephone line. Make sure that you are on the 50-volt scale at least. When the meter is reading properly, the lead to the voltmeter that is marked "common" is the negative lead.

You can check out the Busy Phone before hooking it up to your telephone simply be pressing pushbutton SW1. As soon as it is pressed, you should hear the click of a relay. Do not be alarmed if the LED doesn't light up, as it draws its power from the telephone line and will not work until you have connected the Busy Phone to the line. When you release SW1, you should *not* hear another relay click. You may wait the full 30 minutes to be sure that it is working, or you may simply press the reset switch. When you press SW2, you should hear the relay click once again. This is a sign that it has dropped out and that the circuit is working properly. You are now ready to hook the circuit up to your phone. After you do, press SW1 again and see if the LED comes on. If it does not, chances are that you've hooked it up backwards. Once you get the LED to light, hit the reset button and make sure that it goes out. Now call a friend and ask him to call you right back and see if the telephone is busy. Hang up and push the arm button (SW1). Wait a few minutes and then call your friend back. Don't be suprised when he tells you that he called you back but the line was busy.

Parts List

B1	9-volt battery
C1	330-μF electrolytic capacitor (16 WVDC)
C2	0.01-μF capacitor
D1	light-emitting diode (LED)
IC1	555 IC timer
R1	5.1-megohm, ½-watt resistor
R2	220-ohm, ½-watt resistor
RY1	6-volt (100 mA maximum) dpst relay
SW1	normally open momentary spst pushbutton switch
SW2	momentary dpdt pushbutton switch

9

Crank Call Eliminator

Did you or anyone in your family ever get strange or obscene phone calls? Instead of encouraging these callers by talking to them or even listening to them, let them listen to this device. Not only will it give them an audio shock, but it will let them know that your are quite ready to deal with their intrusions and that all they'll be doing is wasting their time.

Theory of Operation

The heart of this device is the 555 timer IC, which can be hooked up to work as an audio oscillator. Since most of the complicated circuitry required for an oscillator is already built into the 555 IC, only a handful of extra components are needed to turn it into an adjustable oscillator.

The frequency of oscillation is determined by components R1, R2, and C1 according to the following formula:

$$f = 1.44/[(R1 + 2R2)C]$$

where the resistors are in megohms; the capacitor, in microfarads; and the frequency, in hertz. The output of the oscillator is on pin 3 of the integrated circuit. The audio signal produced by the 9-volt battery recommended is quite loud if a 2- to 3-inch speaker is used and can be fed into the telephone acoustically by placing the speaker near the microphone of the handset. A more effective way of transferring the audio energy to the telephone is to add the little circuit shown in Fig. 9.1. The 8-ohm winding of the transformer replaces the speaker in the oscillator circuit, and the other side of the transformer is connected via a 1-μF capacitor to the telephone line. With this direct connection, more energy will be transferred to the telephone line. In either case, the amplitude of the signal is sufficiently loud to cause discomfort to anyone listening to it

Construction and Checkout

If you look at the formula that determines the frequency of the oscillator, you will see that if R2 is chosen to be much larger than R1, then the frequency is inversely proportional to R2 and can be varied over a wide range simply by varying R2. That is the reason a 250-kilohm trimpot is used for R2 — to make it possible to vary the signal produced throughout the audio range. After the circuit is built, R2 should be adjusted

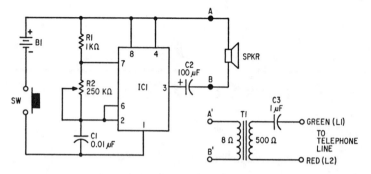

Fig. 9.1 Circuit for crank call eliminator

for about 2 kilohertz because most voice-quality telephone lines are designed to pass signals of up to 3000 Hz, and 2000 Hz is right in the middle of this range. You should find that 2000 Hz will also produce the loudest signal, which is what you are really interested in. The accuracy of the frequency is not really important, and no special equipment for adjustment is needed. Just turn the trimpot for the loudest signal.

If you have some frequency-measuring equipment available, it would be a good idea to check the frequency to make sure that it is not close to 2600 Hz. This is a special frequency that is used by central office equipment, and if you generate a signal of this frequency, you are bound to cause problems with your telephone.

If you are going to use the direct connection scheme, it is suggested that you use a jack/plug to interface your device to the telephone. This will permit you to connect and disconnect the oscillator from the phone line quickly. As an alternative, you might consider using the new modular connectors and purchase an adapter that will allow you to connect two devices to one modular jack (like a Y connector).

The next time you get an unwanted phone call, just give a little press on the pushbutton and see how quickly your offending party hangs up.

Parts List

B1	9-volt battery
C1	0.01-μF capacitor
C2	100-μF electrolytic capacitor (16 WVDC)
C3	1-μF capacitor
IC1	555 IC timer
R1	1-kilohm, $\frac{1}{2}$-watt resistor
R2	250-kilohm potentiometer
SPKR	8-ohm speaker
SW	normally open momentary spst pushbutton switch
T1	8- to 500-ohm audio output transformer

10

TeleTime: Telephone Call Timer

How many times have you stayed on the telephone for an hour or more without even realizing it? Quite a few if you're like most people. The telephone company is of course delighted when you do; it's money in the bank for them. Unfortunately, the money is coming out of your pocket.

You can remedy the situation by using the call length limiter described elsewhere in this book, but that restricts you to a fixed time all the time and does not permit an occasional longer call unless you disable it. With TeleTime, however, you can keep track of the exact length of your conversations and afterwards can even calculate the costs. By keeping records of long distance calls, you may also gain an added bonus. If you discover that you are being overcharged for your phone service, you can make back the cost of this unit in a short time. Does that sound far out? It shouldn't. Sometimes the telephone company inadvertantly charges customers for calls they never made. In fact, many people make a living as telephone consultants and save businesses thousands of dollars a year by tracking down extraneous phone charges.

Overcharges are not all that common, but if you have a record of your calls and their lengths, at least you've got the ammunition to fight with. An additional advantage of the TeleTime is that it can help you break bad telephone habits. With this instant feedback to remind you how much time you are spending on the telephone, you may learn to talk at less length.

Theory of Operation

The heart of the TeleTime is the ubiquitous 555 IC timer, which in this application is connected to operate as an astable multivibrator (oscillator). The key to the correct functioning of the circuit is getting the astable to oscillate at 0.166666 Hz. At this frequency, the timer puts out a pulse on pin 3 once every six seconds.

As can be seen from the block diagram in Fig. 10.1, the output pulses from the 555 are fed into three divide-by-ten counters. These counters are connected to a three-digit display that indicates the elapsed time in tens of

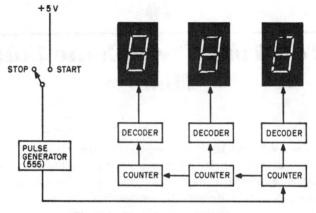

Fig. 10.1   Block diagram of TeleTime

minutes and also in tenths of a minute (one-tenth of a minute is six seconds).

The number of pulses from the timer (tenths of a minute) are counted by IC2, IC3, and IC4 (Fig. 10.2). These ICs are cascaded so that the divide-by-ten output pin of one is connected to the input pin of the next. Thus, IC4 counts tenths of minutes, IC3 counts minutes, and IC2 counts tens of minutes. The output of these counter ICs appears on four pins in the form of BCD (binary-coded decimal) data. These four pins on each counter IC (7490) go high or low according to the count. Since they have values of 1, 2, 4, and 8, for a count of 6, the first pin would have a ZERO logic level, the second and the third a ONE logic level, and the fourth a ZERO logic level $(0 + 2 + 4 + 0 = 6)$.

These BCD outputs are fed to seven-segment decoder/driver ICs (7447) that convert the BCD code to a seven-segment code that will display the digit on the digital display. The seven output lines from each decoder are connected to common-anode LED displays through current-limiting resistors.

The timing interval is started when SW2 is placed in the START position, thereby applying power to the 555 IC oscillator. This, of course, takes place after SW1 has been turned on. At the end of the timing interval, SW2 is switched to the STOP position, removing power from the oscillator but maintaining it to the rest of the circuit. To clear the display and reset it to zero, SW3 must be pressed. This reset switch is a normally closed pushbutton. While closed, it connects the reset pins on IC2, IC3, and IC4 to ground. When the button is pushed, the ground connection is cut, these pins go high, and the counters are thus reset.

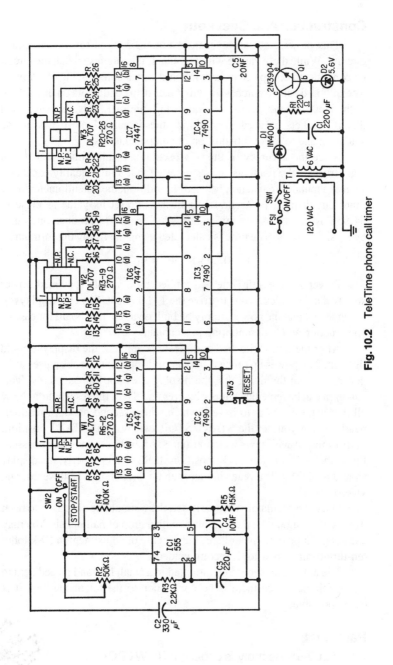

Fig. 10.2 TeleTime phone call timer

Construction and Checkout

This project, as can be seen in Fig. 10.2, uses several integrated circuits. While these can be wired by hand, it is suggested that the circuit be constructed on a piece of vector board with a hole spacing of 0.1 inch. Vector board can be purchased with metal strip conductors on one side or both sides, making point-to-point wiring minimal. It is also suggested that IC sockets be used to minimize the damage done to ICs during soldering and to make substitution of components easy.

Since the accuracy of the 555 oscillator is quite important if you want trustworthy readings, it is suggested that a high quality capacitor, such as a tantalum capacitor, be used. In addition, the potentiometer in the timer circuit must be accurately adjusted. Preset its value as close to 34,955 as you can since this is the value needed to produce one pulse every six seconds. The time is calculated using the following formula:

$$t = 0.693(R2 + 2R3)(C3)$$

Power for the TeleTime is supplied by a simple regulated power supply that provides 5 volts to drive the TTL logic ICs and the displays. It is suggested that the power supply be built on a separate board for ease of construction and later mounting in a cabinet.

After the unit is finished and power is applied, the displays should light up. The power-up process may permit some extraneous pulses to be generated with the result that the display does not indicate zeros. This situation can be remedied simply by pressing SW3, the reset button. Now all that is necessary is to have a watch with a second hand, or readout, nearby when you put the START/STOP switch in the START position. Start timing it with your watch at once. After 30 seconds have passed, throw the switch to the STOP position. If R2 is set properly, the display should read 00.5. If it reads higher than this, R2 should be lowered, and vice versa.

Once the oscillator is properly set, there should be no reason to reset it. Its setting should be fixed in place with a drop of nail polish. You may use a regular potentiometer here, but for ease of adjustment, a 50-kilohm ten-turn trimmer resistor is recommended.

The entire unit should be mounted in a small box and placed next to the telephone. Since there is no connection to the telephone line, it is easily portable.

Parts List

C1   2200-μF electrolytic capacitor (10 WVDC)
C2   330-μF electrolytic capacitor (6 WVDC)

C3	220-μF electrolytic capacitor (6 WVDC)
C4	10-nF capacitor
C5	20-nF capacitor
D1	1N4001 silicon diode
D2	5.6-volt zener diode
FS1	1-amp fuse
IC1	555 timer IC
IC2–IC4	7490 decade counter
IC5–IC7	7447 BCD to seven-segment decoder
Q1	2N3904
R1	120-ohm resistor
R2	50-kilohm, 10-turn trimpot
R3	2.2-kilohm resistor
R4	100-kilohm resistor
R5	15-kilohm resistor
R6–R26	270-ohm resistors
SW1	spst toggle or slide switch
SW2	spdt toggle or slide switch
SW3	spst normally closed pushbutton switch
T1	110- to 6-volt transformer (½-amp secondary)
W1–W3	seven-segment common-anode LED displays (DL707 or equivalent)

11

Melodic Ring Generator

If you find the sudden harsh ringing of the phone unnerving every time that you receive a call and if you would prefer a soothing, lyrical, musical melody instead, then the melodic ring generator is for you. With this simple accessory in place, you will be able to switch between the soft musical signal or the clamoring bell at will.

Theory of Operation

This circuit can be broken down into four sections: the ring detector, the rhythm generator, the note selector, and the output oscillator. The ring detector is a very simple circuit, consisting of a diode bridge, a capacitor, and a relay (see Fig. 11.1).

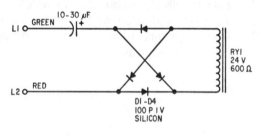

Fig. 11.1   Ring detector circuit

The ringing voltage is about 90 volts at a frequency of about 20 Hz When the ringing signal appears on the line, it passes through C1 and the diodes of the bridge rectifier, applying a dc voltage on the 24-volt relay. As long as a ringing voltage is present, the relay remains closed. When the ringing signal ceases, the relay opens up. This ring detector differs from others used in projects in this book in that the relay contacts do not open and close in sync with the ac ringing signal but close once and stay closed for every time period during which the ring signal is gated on. Once the ring detector is set up, it is an easy matter to produce a customized ring signal with the aid of a programmable tone generator (Fig. 11.2).

The programmable tone generator itself can be broken down into several sections. The rhythm generator consists of an astable multivib-

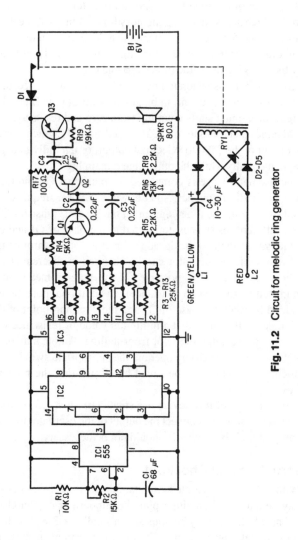

Fig. 11.2 Circuit for melodic ring generator

rator that has been built from a 555 IC timer. Although the frequency is adjustable with the potentiometer, it should be set to less than 1 Hz so that the entire ten-note tune can be played for each ring of the phone.

The notes of the tune to be played are determined by ten variable resistors, which act as an analog memory for the tune. Unlike the musical synthesizer used for one of the telephone hold circuits in this book, each note of this electronic music generator is individually programmable simply by changing the value of the variable resistor.

After the pulse is generated by the 555, it is fed to a decade counter (IC2) at pin 14, and the count, in binary, appears at pins 12, 9, 8, and 11. As wired, the counter starts counting from zero and resets to zero after a count of nine. The outputs of IC2 are fed to IC3, which converts the binary output of the 7490 to a decimal output, or ten individual lines that go low sequentially while the other nine remain high.

The oscillator stage is comprised of transistors Q1 and Q2, which produce an audio-frequency square wave whose frequency can be changed by varying the time constant at the base of transistor Q1. Thus, with ten different resistances connected to the base of Q1 and each one individually selectable as the decade counter advances through its cycle, the frequency of the oscillator can be changed in discrete steps to ten different tones. Since it is possible for the potentiometers to be at a minimum setting (almost no resistance), an additional potentiometer has been added between the base of Q1 and the common point of the other potentiometers. It assures that a minimum value of resistance will be in the circuit to prevent the oscillator from stalling. A note may be blanked out (that is, a rest inserted) by opening the appropriate switch. If no rests are anticipated, money can be saved by eliminating the note-selecting switches.

The last stage of this device is a small amplifier, Q3. If you wish a higher volume, you may connect the output of the oscillator to an external audio amplifier. This, however, should not be necessary.

Construction and Checkout

With only three integrated circuits, this device is fairly easy to build and may be attempted on either plain breadboard or copper-clad vector board. The whole unit may be housed in a small 5 × 3 × 2-inch minibox with an external power supply. The most convenient of these is a calculator or cassette recorder supply, which are easily available at low cost. Make sure that it is a 6-volt supply. The diode in series with it will bring this voltage down to just a little less than 5.5 volts, a value near the upper limit of permissible voltage for the TTL circuits used here.

Once the music generator has been built, it can be tested by simply shorting out the relay contact that connects it to the power supply. The potentiometers should then be adjusted to play the tune you want. For adjustment purposes, you may want to speed up the rhythm generator so that you can set the tune quickly. After doing so, don't forget to slow the 555 astable back down.

Once it has been programmed, connect the diode bridge part of the device to the telephone line (in parallel). Next, have someone call you. If the melody generated does not have enough time to finish for each ring of the phone, simply speed up the 555 astable a bit.

Parts List

B1	6-volt battery
C1	68-μF electrolytic capacitor (10 WVDC)
C2, C3	0.22-μF capacitor
C4	10- to 30-μF electrolytic capacitor (150 WVDC)
D1	IN4001 silicon diode
D2–D5	diode bridge (150 PIV)
IC1	555 timer IC
IC2	7490 decade counter
IC3	74141 BCD to decimal decoder
Q1–Q3	2N3406 pnp transistor or equivalent
R1	10-kilohm resistor
R2	15-kilohm resistor
R3–R13	25-kilohm trimpots
R14	5-kilohm trimpot
R15, R18	2.2-kilohm resistor
R16	13-kilohm resistor
R17	100-ohm resistor
R19	39-kilohm resistor
RY1	24-volt, 600-ohm spst normally open relay
SPKR	80-ohm speaker (8-ohm speaker can be used with less efficiency)

12

Two-Station Telephone Intercom

Do you need a private intercom between two points in your house, or are you looking for a fun toy that will keep your kids busy for hours? If so, then dig out those spare telephones or get a pair for under $20 from your nearest flea market or electronic parts store. Next, add a handful of electronic parts and you now have a two-station intercom, in which each station automatically rings the other the moment that a telephone handset is picked up.

Theory of Operation

Most telephones have a handset that contains an electromagnetic earphone and a microphone containing carbon granules. Sounds picked up by the microphone are variations in air pressure. These variations cause a diaphragm to vibrate. The vibrations are then applied to the carbon granules, which act as a variable resistor. When a dc voltage is connected to the carbon microphone, the changes in resistance produced by the sound waves become changes in current that can be transmitted over an electrical circuit.

To retrieve sound from this varying current signal, the signal is applied to the electromagnet of the earphone. This varying electrical signal causes a varying magnetic field to be developed. The varying magnetic field causes a diaphragm in the earphone to vibrate at a varying rate, reproducing the original sound.

Thus, if we take two telephones and connect them together with a battery in between them, we'll be able to talk from one telephone to the next (see Fig. 12.1).

Although a telephone's main job is to provide a means by which conversation can be carried over a considerable distance, a second and almost as important function is to alert someone that someone else wishes to speak with him. This is done by ringing a bell located in the telephone. In order to ring this bell, the telephone company sends out an alternating

---

* Adapted from P. Stark, "Private Telephone: Simple Two-Station Intercom," *Modern Electronics,* July 1978, p. 32.

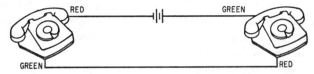

Fig. 12.1 Simplified telephone intercom

current signal with a frequency of about 20 Hz and an amplitude of about 100 volts.

Since generating this type of signal can be bothersome, we can eliminate the need for it altogether by using a low voltage bell or buzzer. The small piezoelectric buzzers that are available at electronic parts stores are recommended. Ordinarily, this would require the use of another one or two wires, just for signalling. But with the addition of a one-transistor electronic switch, it is possible to use the same pair of wires that are used for talking over the telephone.

The actual switching is performed by transistors Q1 and Q2. In order to make the transistor conduct and the buzzer sound, a dc voltage of at least 1 volt must be applied to D1 or D2. Which transistor turns on depends on which telephone receiver is lifted off the hook first. If one telephone handset is lifted off the hook, the low resistance of the telephone is placed in parallel with the switch for its buzzer circuit. As a result, a very low voltage drop appears across the telephone, and thus a very low voltage is applied to the diode and the transistor cannot turn on.

Since the telephone that is being called, however, still has its receiver on the hook, the hook switch is open and the telephone is not connected across the line. Since there is no low resistance in parallel with the switch for this buzzer, this circuit sees almost the full voltage of the battery (1.5 volts) that is powering the talk circuit of the two telephones. As a result, the transistor turns on and the buzzer sounds. When the phone is answered, the voltage drops to less than 1 volt; and the buzzer turns off. When both phones are hung up, the voltage applied to each phone is equal to about 0.75 volt, or one-half the battery voltage. Since this is less than 1 volt, neither buzzer will sound.

Construction and Checkout

Since the electronic circuitry involved is so simple, any type of construction is permissible. The simplest is probably to hand wire the components in a small piece of vector board or breadboard. The sounding device can be a bell or a buzzer. As mentioned earlier, piezoelectric buzzers are recommended because they can be operated from low voltages with little current. If you choose, however, to use a bell or buzzer that has an electromagnet in it (for example, an ordinary doorbell) then you should also use diodes D3 and D4. These eliminate the inductive kick, or back emf, produced by the electromagnet, a kick that could otherwise destroy the transistor.

Since the circuit is physically small, it can fit inside the telephone along with the buzzer. Each buzzer circuit will need its own power source, which can vary from 1.5 to 9 volts, depending on the amount needed for the particular buzzer you have chosen.

Once you have built the intercom, connect it as shown in Fig. 12.2. At this point neither of the telephones should be buzzing (both receivers are on the hook). If one of them is buzzing, the hook switch is probably shorted. Now lift one receiver off the hook. The other telephone should start buzzing right away. If it doesn't, check to make sure that the telephones are properly wired. Only the red and green wires are used. The red of one telephone goes to the green of the other, and the remaining green and red wires go to the positive and negative sides of the 1.5-volt battery, respectively. You can check the individual phones simply by applying 1.5 volts to the green and red wires (green is positive). If a phone still doesn't buzz, it is possible that you have connected your buzzer circuit after the hook switch; as a result, it will not see the 1.5 volts until after the phone has been taken off the hook. This should be corrected.

If one phone buzzes when the other is taken off the hook, answer it. At this point, neither of the buzzer circuits should be sounding. If one of them is, it is most likely due to a mismatch in the resistance of the individual telephone units. Different types of telephones have different resistances, and the resistance of the phone that is buzzing is higher than that of the other. To compensate, you can add a resistor in series with the one in the telephone that is not buzzing so that its resistance equals the resistance of the phone that is buzzing. Remember that when you measure the resistance of a telephone, the handset must be off the hook, and the battery connected between the two phones must be disconnected so that you don't damage the ohmmeter.

Parts List

B1–B2	1.5- to 9-volt battery, depending on buzzer used
B3	1.5-volt battery only
BZ1–BZ2	Buzzers or bells
C1–C4	0.1-μF capacitors
D1–D4	IN4001 silicon diodes
Q1–Q2	2N3904 npn transistors
R1–R2	100-kilohm, ½-watt resistors
R3–R4	100-ohm, ½-watt resistors

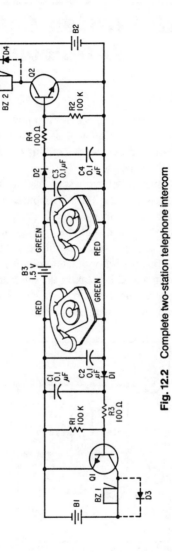

Fig. 12.2 Complete two-station telephone intercom

13

Multi-Station Telephone Intercom*

While the previous project showed how to build an intercom with two telephones, this section will show how to make a multi-station intercom. Unlike the previous project, this one does not require the entire telephone, just the handset.

A large number of intercom stations (Fig. 13.1) can be tied together, party-line style, with only three interconnecting wires. If a common-ground system is used, only a single twisted pair of wires is needed.

For redundancy, each station is individually powered by a 1.5-volt battery (if "D" cells are used, they should last about a year), but if large distances are to be covered or greater volume is desired, a 3-volt battery should be substituted. If a break in the interconnecting wires develops, the redundant power supplies permit stations on either side of the break to continue to communicate with each other.

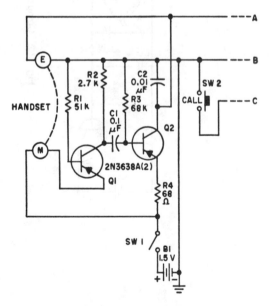

Fig. 13.1 Single station

42

Theory of Operation

All intercom stations are connected in parallel, and the entire system uses one central oscillator to provide signaling. The carbon microphone of the standard telephone handset at each station is connected to a common-base amplifier, which in turn is connected to a high-gain common-emitter amplifier. The common-emitter amplifier drives the intercom line. All handset earpieces are connected in parallel across the line.

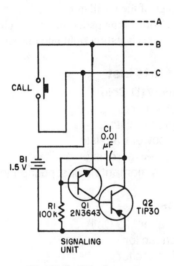

Fig. 13.2 Signaling unit

The signaling circuit that is used with the intercom (Fig. 13.2) is a two-transistor complementary astable multivibrator. The advantage of this particular circuit is that it requires only four components and needs little power. As with the individual stations, the signaling oscillator also has its own power source, a 1.5-volt battery. Since it is connected across the line to all earphones, all stations will sound when the buzzer is activated. To get a particular station, therefore, a system of buzz codes must be used.

Construction and Checkout

Construction of the intercom is quite straightforward since no critical wiring is involved. Since only a few parts are needed for each station,

---

* A.M. Hudor Jr., "Party-Line Intercom System Needs Only Three Wires," *Electronic Design* 16, Aug. 2, 1976, p. 76.

the circuit can be built on a perforated board and housed in a small minibox. One circuit is needed for each station. The signaling circuit can be built in its own special box or included in one of the stations. All stations are masters.

To check out the system, simply connect all the A wires of each station and the signaling circuit together. Then connect all the B wires and C wires. You're now ready to try the system out. Press any one of the call buttons to see if all the earphones buzz. If you have many stations, the buzzing may not be very loud. The volume may be increased by increasing the battery voltage of the oscillator.

One final note: With a little patience, you should be able to fit all the electronics needed for each station right into the headset case.

Signaling Unit Parts List

B1	1.5-volt battery (D Cell)
C1	0.01-μF capacitor
Q1	2N3904 npn transistor
Q2	TIP 30 pnp power transistor
R1	100-kilohm, ½-watt resistor
SW	normally open momentary spst pushbutton switch

Station Parts List

B1	1.5-volt battery (D Cell)
C1	0.1-μF capacitor
C2	0.01-μF capacitor
Q1-Q2	2N3906 pnp transistors
R1	51-ohm, ½-watt resistor
R2	2.7-kilohm, ½-watt resistor
R3	68-kilohm, ½-watt resistor
R4	68-ohm, ½-watt resistor
SW1	spst slide or toggle switch

14

Ringing Telephone Intercom *

Yet another telephone intercom using surplus phones can be built. This one, however, unlike the previous two, does not need an external signaling circuit. Signaling is carried out with the telephone's own bell.

As mentioned previously, quite a high-voltage ac source is usually needed to provide the ring signal that activates the bell. This drawback can be overcome if an additional component — a low frequency choke — is added to the circuit and, more particularly, if the European phones now available in surplus stores are used.

How It Works

The connection of the telephones in this circuit is slightly different from the connection in the previous project. Here they are connected in parallel with one another and the voltage source. The previous project connected them in series.

The two parallel-connected telephones are fed with direct current through a low-frequency choke, as shown in Fig. 14.1. When a caller lifts up his handset, the cradle switch contacts close, energizing the carbon-microphone transmitter. To signal the other party, the caller must dial a number (preferably a zero). When the number is dialed, the normally open dial contacts close, causing a short circuit across the line. This is the key reason why European phones work best. From the telephone schematic in Fig. 14.2, it is clear that the normally opened dial contact will short

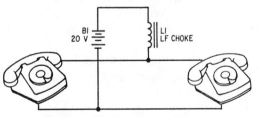

Fig. 14.1 Ringing telephone intercom

*Adapted from J.T. Lawrence, "Electronic Home Telephone Exchange," *Practical Wireless*, Feb. 1975, p. 892.

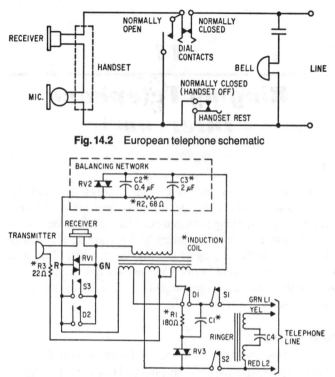

Fig. 14.2 European telephone schematic

Fig. 14.3 Standard American (500-type) telephone schematic

out the telephone line when it closes. If you compare Fig. 14.2 with the schematic of a standard Bell telephone in Fig. 14.3, you can see that the short will not occur.

The short circuit across the line causes a heavy current to flow through the low-frequency choke. When the dial returns to its normal resting position, the dial contacts open and close ten times (for a zero), producing a substantial pulsing voltage across the line. When this voltage is passed through the bell and the series capacitor, the phone rings. When the dial comes to rest, the normally open contacts open, and current again flows through the transmitter. When the called person lifts up his handset, his transmitter also becomes energized.

During conversation, variations in transmitter resistance will cause speech voltages to be developed across the line, and incoming speech voltages will operate the receiver. The low-frequency choke provides a direct-current path that energizes the transmitters while presenting a high ac impedance to the speech signals.

Construction and Installation

Construction of the intercom is pretty simple. All you have to do is connect two telephones in parallel with a 20- to 24-V battery (or series combination of batteries) that is in series with a 350-ohm, 3-henry inductor.

Since the pulsing rate is different from the frequency normally used for ringing the bell, it may be necessary to readjust the bell to improve the tone. To do so, the base of the telephone must be removed. The bell fixing hole is off center, and loosening the fixing screw will allow the bell to be rotated and then tightened to give the best sound.

Telephone Modification

This project requires a minor modification of the standard 500-type telephone since it depends on the telephone's being shorted out by one of the dial contacts whenever the dial is not in the rest position. While many of the foreign phones that have come into this country do short out in this way, it is not a feature of American phones. Instead, the normally open switch located on the back of the American telephone dial is connected to the telephone network block and is used to short out the receiver during dialing so that the dialing clicks won't be heard.

To make the modification, simply disconnect the two wires from the GN and R terminals on the block and connect them as shown in Fig. 14.4. Make sure that you have the right wires. They come from the normally open switch on the dial that closes and stays closed as long as the dial has been rotated from its rest position. In Bell desk phones, the wires are white.

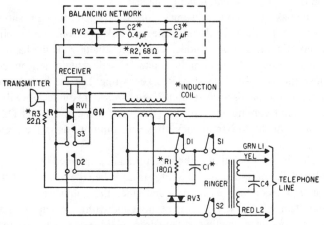

Fig. 14.4   Telephone modification

Parts List

B1   20- to 24-volt battery
L1   3-henry, 350-ohm choke

15

Home Telephone Exchange *

If you want to run an intercom between more than two stations, the intercoms described on pages 38 and 45 are not too useful. If privacy is not a key factor, you can use the circuit described on page 42. However, if you want a true telephone-like intercom, then this project is for you. It allows you to have as many as nine separate telephones connected together, each of which can be individually called. Only the called phone rings, allowing for increased privacy.

Theory of Operation

Basically, this circuit works in much the same way as the previous one. The power source is in series with an inductor, which is then connected in parallel with the various telephones. This system, however, incorporates a method of routing the message from one telephone to the telephone called. Most systems use a stepping relay for this purpose, but it is both noisy and slow. Here, diodes are used for speech routing, and reed relays are used to ring the desired phone. A phone is selected by dialing any number from 1 to 9, except, of course, the number of the phone from which the call is being made.

The basic technique of using diodes for speech routing is shown in Fig. 15.1. With their handsets in the on-hook position, the telephones draw no current, and both diodes are nonconducting. If the handset of station 1 is lifted, D1 will conduct, and D2 will remain nonconducting because the associated resistor, R2, keeps line 2 at the battery supply voltage. If the handset of station 2 is now lifted, diode D2 will also conduct, and communication can take place through D1 and D2; the impedance of the low-frequency choke is common, as in the previous project.

The ringing operation in the system can be explained by looking at a simplified diagram such as that in Fig. 15.2. When the handset is lifted at station 1, D1 conducts as before, and since the voltage on the line is greater than +7 volts, D3 remains nonconducting. When the dial on station 1 is operated, the line voltage drops and D3 conducts, passing the dialing pulses into the counting and ringing circuit.

*J.T. Lawrence, "Electronic Home Telephone Exchange," *Practical Wireless*, Feb. 1975, p. 892.

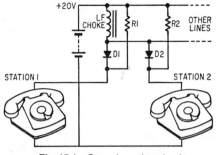

Fig. 15.1 Speech routing circuit

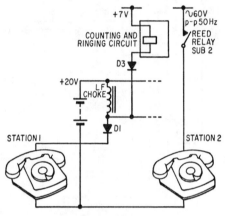

Fig. 15.2 Ringing circuit

If the caller, station 1, has dialed the number 2, the counting and ringing circuit will energize the reed relay for station 2 and cause a 60-volt peak-to-peak ac signal to be connected to line 2, thus ringing the bell at station 2. As soon as the handset at station 2 is lifted, diode D2 (omitted from Fig. 15.2 for clarity) conducts, allowing communication as described previously. A separate reed relay is required for each station.

The full circuit appears in Figs. 15.3, 15.4, and 15.5. The circuit operation begins when the caller lifts a handset. Assume, as before, that the caller is station 1. When the handset is lifted, current will flow from the +20-volt supply through L1, D13, and the station 1 telephone to the common line (see Fig. 15.4). This telephone is now energized, and the voltage across the line is approximately 8 to 16 volts, depending on the total resistance of the telephone.

Assume that station 1 dials station 3. On the backward rotation of the dial, the line is shorted by the dialing contacts, and the voltage across the line is zero. Current now flows through relay A, D12, and D13, thus energizing relay A (Fig. 15.4). When station 1 releases the dial, it returns

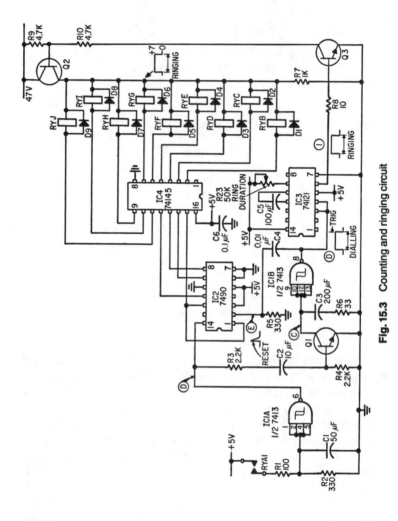

Fig. 15.3 Counting and ringing circuit

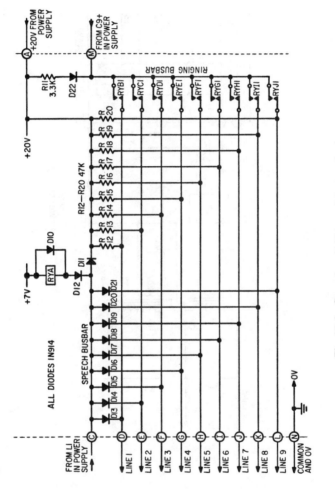

Fig. 15.4 Speech switching and ringing circuit

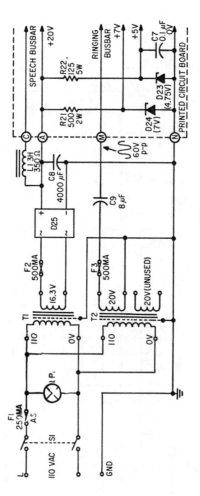

Fig. 15.5   Power supply

to rest, whereupon contacts open and short the line three times, finally coming to rest with the telephone energized as previously. The waveforms are shown in Fig. 15.6.

The contacts of relay A connect R1 to the +5-volt supply, Fig. 15.3, and a logic 1 level is applied to the Schmitt trigger gate IC1a. A Schmitt trigger is used in this position so that the input signal can be smoothed by R1, C1 to eliminate spurious signals caused by contact bounce and yet provide a fast-switching, clean output suitable for driving the logic circuits. The output of IC1a, containing the dialed information, now follows two routes, to Q1 and IC2.

In order to detect the completion of a group of dialing pulses and to produce a signal at the end of this group, it is necessary to have an electronic circuit equivalent to a "slugged" relay. This is formed by Q1 and IC1b. The first of the dialing pulses turns Q1 hard on, discharging C3 and causing the inputs of IC1b to be at logic 0 and thus the output of IC1b to be at logic 1. This positive transition is differentiated by C4 and R5 to produce a positive pulse to logic 1, which resets the 7490 decade counter to zero.

Between each dialing pulse, C3 begins to recharge from the input current of IC1b but does not reach logic 1 level because the next dialing pulse discharges it again. However, after the last dialing pulse, the input to IC1b rises to logic 1 and the output flips to logic 0. The other route for the dialing pulses is direct to the input of the 7490 decade counter, IC2. The counter counts on the negative-going edge of each dialing pulse and in our example would count 3 (binary 0011).

It will be noted from the waveforms that a negative step (to logic 0) occurs when the dial is first rotated, and this is counted by the decade counter, but as the counter is reset to zero on the leading edge of the first true dialing pulse, it does not cause a dialing error.

The binary-coded decimal output from IC2 is fed directly to the 74145 decoder-driver, IC4, where, in our example, output 3 would be clamped to the 0-V line, and all other outputs would be nonconducting.

Returning now to IC1b, at the end of a group of dialing pulses, the output flips from logic 1 to logic 0. This negative transition triggers the 74121 monostable IC3, which produces a logic 1 output pulse having a duration set by R23 (1 to 5 seconds approx.). The positive pulse from IC3 turns on Q3 and Q2, thus energizing only relay D, as this is connected to ground through the No. 3 output of IC4.

The reed relay D connects an ac ringing signal of 60 volts peak-peak from T2 to line 3, which rings the bell in station 3 telephone for a duration set by VR1. When station 3 lifts his handset, current flows through D15, thus connecting both telephones to the speech bus bar and permitting

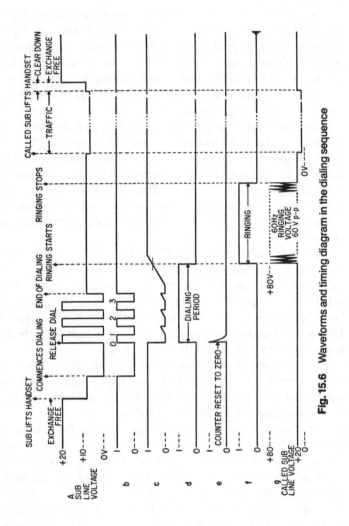

Fig.15.6 Waveforms and timing diagram in the dialing sequence

two-way conversation. At the completion of the call, the caller and the called replace their handsets, causing D13 and D15 to disconnect. No other "clearing down" is required. The decade counter is reset to zero at the commencement of the next call.

The 60-Hz ringing voltage from T2 is ac-coupled to the ringing bus bar by C9, and the negative peaks cause D22 to conduct, charging C9 and thus preventing the negative half cycles from swinging below +20 volts. This ensures that when the ringing signal from the ringing bus is connected to a subscriber it will not cause the associated speech routing diode to conduct and will prevent the ringing signal from appearing on the speech bus.

The inductance of the low-frequency choke, L1, is not critical, but the resistance should be approximately 350 ohms. If a choke of the correct resistance is not available, one having a lower resistance may be used, with a suitable resistor connected in series to raise the total resistance to the correct value.

The power supply for the telephone exchange is in operation continuously and consumes approximately 3 watts. Transformer T1, bridge rectifier D25, and C8 provide the +20-volt supply for powering the telephones and this is also the source of the +5-volt and +7-volt supplies, which are regulated by zener/breakdown diodes D23 and D24, respectively (Fig. 15.5).

Transformer T2 provides the ringing voltage for the ringing bus. Both transformers are adequately fused for safe unattended operation

All the circuitry with the exception of the power-supply components is built on an 8 × 4¾-in. printed circuit board (Figs. 15.7 and 15.8). The power-supply components are mounted on a similarly sized metal plate, and the complete unit is housed in a 5 × 5-in. case that is 8½ in. long.

Connection and Operation

Each telephone is connected between the appropriate line terminal and the common line terminal. Dialing is carried out in the usual way, and VR1 may be adjusted to give the required length of ring. The telephone bells may be adjusted for best ringing, as described previously.

The system has the advantage that when the two subscribers are in communication a further subscriber may be dialed and join in the conversation. Two telephones may be connected in parallel on one line if required.

Telephone Modification

As with the previous project, this telephone exchange was originally built to accommodate some of the inexpensive foreign-made telephones

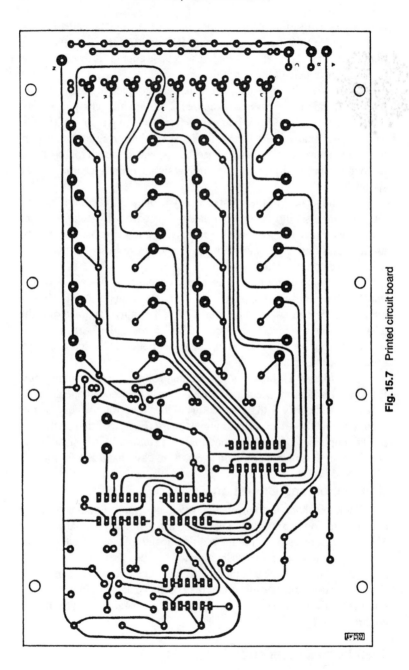

Fig. 15.7 Printed circuit board

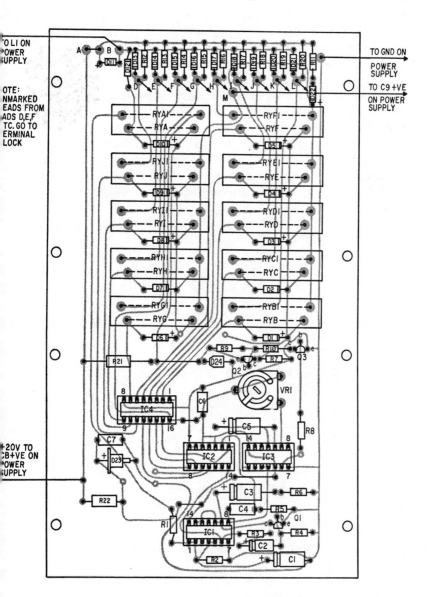

Fig. 15.8 Parts layout on circuit board

that surface in surplus electronics stores. The main difference between these phones and standard American phones is the contact on the back of the dial that shorts the phone during the dialing operation.If you don't want to use a surplus phone, you can use a standard American phone simply by opening it up and connecting directly to the contacts on the back of the dial. In the standard Bell system 500-type telephone, two white wires from this switch go to terminals GN and R on the network block. Since this normally open switch closes when the dial is moved from the rest position, and stays closed until the dial returns to the rest position, it is used as a mute switch to eliminate the clicking sound made in the earphone when the dial opens and closes its contacts. In our application, these two wires are disconnected from the network block and connected as shown in Fig. 15.9. Because of the modifications made to the telephone and the way it is connected, it is not recommended that this system be hooked up to the outside telephone system.

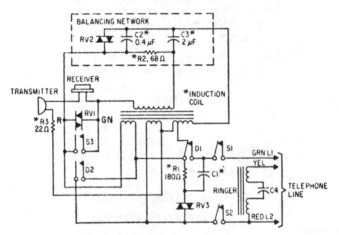

Fig. 15.9 Telephone modification

Parts List

C1	50-μF electrolytic capacitor (6 WVDC)
C2	10-μF electrolytic capacitor (6 WVDC)
C3	200-μF electrolytic capacitor (6 WVDC)
C4	0.01-μF capacitor
C5	100-μF electrolytic capacitor (6 WVDC)
C6-C7	0.1-μF capacitor
C8	4000-μF electrolytic capacitor (40 WVDC)
C9	8-μF electrolytic capacitor (250 WVDC)
D1–D22	1N914 silicon diodes

D23	4.75-volt zener diode
D24	7-volt zener diode
D25	1-amp, 50-PIV bridge rectifier
F1	¼-amp fuse
F2–F3	½-amp fuses
IC1	7413 dual four-input Schmitt triggers
IC2	7490 decade counter
IC3	74121 monostable multivibrator
IC4	74145 BCD decoder/driver
L1	3-henry, 350-ohm low-frequency choke
Q1, Q3	2N3904 npn transistors
Q2	2N3906 pnp transistor
R1	100-ohm resistor
R2, R5	330-ohm resistor
R3, R4	2.2-kilohm resistor
R6	33-ohm resistor
R7, R8	1-kilohm resistor
R9, R10	4.7-kilohm resistor
R11	3.3-kilohm resistor
R12–R20	47-kilohm resistors
R21	500-ohm, 2-watt wirewound resistor
R22	250-ohm, 5-watt wirewound resistor
R23	50-kilohm trimpot
SW1	dpst toggle switch
T1	110- to 16.3 VAC transformer (½-amp secondary)
T2	110- to dual 20-VAC transformer (½-amp secondaries)

16

All Solid-State Telephone Exchange*

For those of you who want to reduce the size of your projects and eliminate as many electromechanical devices as possible, this version of the home telephone exchange is for you. Moreover, it will probably cost less money to build this exchange than the previous one because the ten reed relays have been eliminated along with the shunt diodes and a few other components. However, two additional low-cost integrated circuits are needed, a 7413 and a 7408. Also, a 74141 decoder IC replaces the 74145 previously used.

Before building this project, it is suggested that you read the preceding section, which discusses the home telephone exchange requiring reed relays, explains the basic operating theory for the exchange, and details the modifications of the telephone instrument that are needed. Furthermore, reference will be made throughout this project to the previous one.

The +7-volt power supply needed to power the reed relays in the previous project is no longer sufficient. This project requires a 60-volt supply. It is easily obtained because the ringing supply required in the former project is no longer necessary, and the same transformer can be used to produce the required 60 volts (see Fig. 16.1).

As far as counting and the generation of the ring pulse are concerned, this circuit operates identically to the previous one. The only difference is that a transistor, Q1, is used to detect the dial pulse instead of relay A.

Theory of Operation

The major difference between the circuitry of this project and that of the previous one is the way the selected line is rung. The binary output of IC2 is fed to the decoder, IC4, through AND gates (Fig. 16.2). IC5b is normally low, and until it goes high, the decoder will not operate. At the end of the dial period, the monostable IC3 goes high for the duration of the ring pulse. IC5a is enabled and produces a pulse train with a frequency

*G.M. Rossetti, "A Solid-State Version Home Telephone Exchange," *Practical Wireless*, March 1976, p. 945.

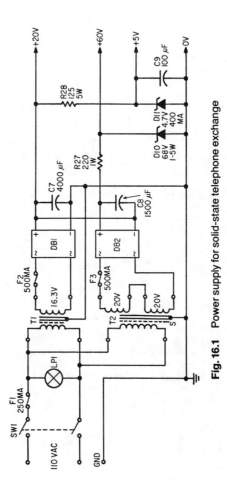

Fig. 16.1 Power supply for solid-state telephone exchange

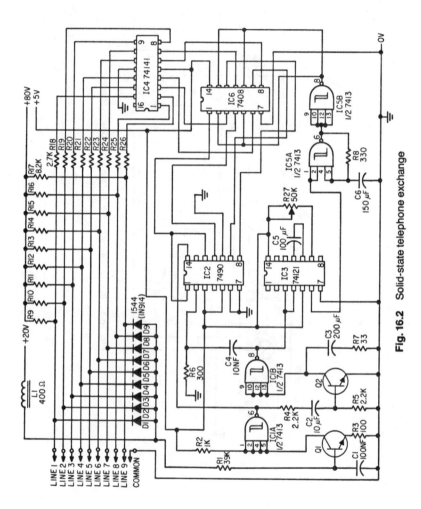

Fig. 16.2 Solid-state telephone exchange

of 16 Hz. This is inverted by IC5b, and the pulse enables the AND gate of IC6. The decoder is thus pulsed on and off at a frequency of 16 Hz.

When IC4 is in the ON mode, it decodes the binary output from IC2, and the selected decimal output goes low. This output is now connected to the called telephone through 2.7-kilohm resistors. Normally, the telephone line is at a potential of 60 volts, but when the decoder operates in response to the ring pulse, this drops to about 15 volts. The result is that a 45-volt ac potential is superimposed on the telephone line at a frequency of 16 Hz and causes the bell to ring.

There are two disadvantages of this circuit when it is compared to the previous one, neither of which is especially significant in home applications. To begin with, its transmission efficiency is slightly lowered by the shunting effect of the 8.2-kilohm resistors. Second, the bell ringing is not as loud nor as shrill as it would be with a 60-volt ac signal.

Construction and Checkout

Since this project uses a few integrated circuits, the best way to build it is to use a breadboard or vector board. Neither construction nor parts placement is critical. If several units are to be built, it would be best to use a printed circuit board to minimize construction time and increase the accuracy of the connections. The unit can be easily checked out simply by connecting two telephones to any two lines and dialing one with the other. If the bell ring is too soft, it may be possible to make it louder by adjusting the knob underneath the telephone (American phones).

Parts List

C1, C9	100-μF capacitors
C2	10-μF electrolytic capacitor (6 WVDC)
C3	200-μF electrolytic capacitor (6 WVDC)
C4	10-μF capacitor
C5	100-μF electrolytic capacitor (6 WVDC)
C6	150-μF electrolytic capacitor (6 WVDC)
C7	4000-μF electrolytic capacitor (40 WVDC)
C8	1500-μF electrolytic capacitor (150 WVDC)
D1–D9	1N914 silicon diodes
D10	68-volt, 1.3-watt zener diode
D11	4.7-volt, 400-mW zener diode
DB1, DB2	1-amp diode bridges (50 PIV)
F1	¼-amp fuse
F2, F3	½-amp fuses

IC1, IC5	7413 dual four-input Schmitt triggers
IC2	7490 decade counter
IC3	74121 monostable multivibrator
IC4	74141 decoder/driver
IC6	7408 quad two-input AND gate
L1	350- to 400-ohm, 3-henry choke
Q1, Q2	2N3904 npn transistors
R1	39-kilohm resistor
R2	1-kilohm resistor
R3	100-ohm resistor
R4, R5	2.2-kilohm resistor
R6, R8	330-ohm resistor
R7	33-ohm resistor
R9–R17	8.2-kilohm resistor
R18–R26	2.7-kilohm resistor
R27	220-ohm, 1-watt resistor
R28	125-ohm, 5-watt resistor
R29	50-kilohm trimpot
SW1	dpst toggle switch
T1	110- to 16.3-VAC transformer (½-amp secondary)
T2	110- to dual 20-VAC transformer (½-amp secondaries)

17

Automatic Telephone Dialer with Display *

Touch-Tone dialing is great! A zero can be dialed just as quickly as a one, and you don't have to watch all those holes go by! What's that — your telephone office doesn't offer Touch-Tone? Well, cheer up! You can build this dial converter and go Touch-Tone two better: The number you are dialing will be displayed by a LED read-out, and you can use the converter as a calculator when it isn't needed for dialing.

The dial converter accepts and displays a keyboard-entered telephone number (up to eight digits). On command, this number is converted digit by digit (most significant digit first) into corresponding dial pulses, thus dialing a number. Dialing can be repeated by retaining the number in memory and pressing the dial switch again. The system is based on an IC that can be used as a calculator when not dialing. All telephone interfacing is performed with high-isolation relays to avoid any telephone network damage from stray voltages caused by failures.

Operation

The converter is based on a calculator using the National Semiconductor MM5738 calculator IC and DM8864 digit driver. Twenty-two connections are made to the calculator — one for each of the eight digits and two for each of the segments. Calculator operation is not affected.

A telephone number is entered as it appears in the book — most significant digit first. The number is displayed on the eight-digit LED read-out. The telephone receiver is taken off the hook and the dial button is pressed.

The dial button loads a one into the first position of digit register IC12 (Fig. 17.1) at the same time that overflow register IC11-a is cleared. Dial pulse oscillators IC7-c and IC9-d are enabled and begin to output 10- to 12-Hz pulses. Dial relay RY1 opens and closes the telephone circuit at this rate and commences "dialing" a digit. Counter IC10 accepts these pulses and synchronously decodes them to 7-segment format. The seg-

*Reprinted from *Radio-Electronics,* November 1976, ©Gernsback Publications Inc., 1976.

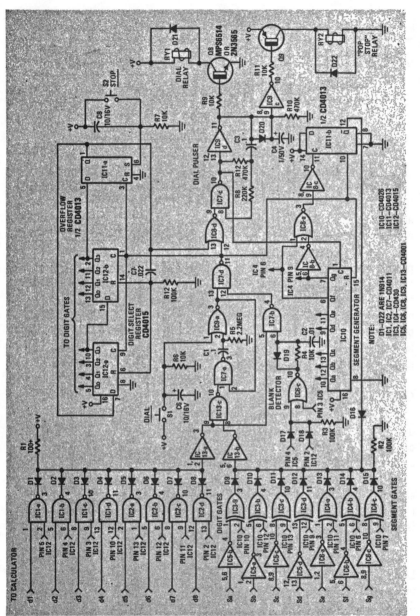

Fig. 17.1   Telephone dial converter (connections marked d_1, s_1, etc., are connections to the calculator—see Fig. 17.2)

ment outputs from the calculator chip and from IC12 are compared in exclusive-OR gates IC3 and IC4 for coincidence. When all segment outputs agree, all exclusive-OR outputs are "zeros" and are inverted and applied to the input of IC13-c. Gates IC1 and IC2 serve to demultiplex the read-out information.

In the calculator, the read-out is multiplexed by having common segment-lines for all digits and presenting segment information for each digit in turn as they are sequentially enabled by the digit drivers. This is done so very rapidly that the human eye cannot register the change and sees a continuous display. Gates IC1 and IC2 serve to select a single digit regardless of the scanning. The output of the exclusive-OR gates may register coincidence several times as the digits are scanned; however, ANDing the outputs of the digit gates and the exclusive-OR gates insures that only the digit selected by the shift register is detected. When the output of both the segment coincidence gates and the selected digit gate are at a logical-1 level, the dial pulser is stopped, IC10 is reset, and an interdigit pause is activated. The interdigit pause serves to identify the end of a digit to the telephone central office.

Since the number of pulses dialed corresponds to the number displayed on the read-out, we have just "dialed" the first number. At the end of the interdigit pause, the digit register is advanced, the next most significant digit is selected, and the dial pulser is again enabled.

This process continues until the least significant digit has been dialed. The "one" that was shifted across the digit register falls into the overflow register, disabling the dial pulser and hence the entire system. The telephone number you selected has now been dialed.

Special Considerations

Dialing a Zero   When the number to be dialed is a zero, we have complications, since in seven-segment language a zero is 0 and in dial language a zero is 10. Thus we have to inhibit the coincidence gates on the initial zero. Initially, IC11-b is reset, and the Q output (at a logic-1 level) prevents the coincidence gates from switching. As soon as IC10 goes from zero to one, IC11-b is clocked to a one via gates IC8-a and IC8-b (which detect the presence of a zero in the segment output of IC10), thus removing the inhibit from the coincidence gates. The next time IC10 reaches zero, the normal end-of-digit is signaled, and since IC10 has stepped only ten times, a 10 has been "dialed."

Skipping Blank Digits   Whenever less than eight digits are entered in the calculator display (which will be most of the time), we must somehow tell the digit register to advance until a valid digit is present. Otherwise the counter and calculator segments would never agree, and the dial pulser would dial forever.

Blank detector gates IC6-c and IC7-b sample segments that are always lighted as long as a number is displayed. The absence of both segments is interpreted as a blank. This output is ANDed with the digit gates and clocks IC12 to the next digit position. This process repeats very rapidly until a nonblank digit position is found and dialing proceeds normally. This skip effect occurs so rapidly that the dial relay never has a chance to operate!

Resistor R4 and capacitor C2 filter out the very narrow pulses that occur in the calculator during the blanking period between each digit. Diode D18 inhibits the blank detector when a digit one is selected. (The calculator scans digits from right to left, but the dial converter reads out from the opposite direction.) When we get to the last digit — calculator first digit — the end-of-word blanking pulse combines with the selection of digit one and immediately clocks IC12 into overflow. Since there never can be a blank in digit-one position — the cleared calculator always displays a zero — it is safe to allow the pulser to operate always in this position.

Calculator Converter Interface

For the calculator to display a segment, the segment driver in the calculator chip raises the anodes of all the selected segments toward V_{DD}; the selected digit driver pulls the cathodes of all the segments toward ground, and current flows through the selected segment, causing it to light.

To interface with the calculator chip, we have to obtain both digit and segment information. Digit interfacing is OK; the levels are compatible with CMOS thresholds. The segments, however, show very little usable voltage change from on to off. There is, of course, plenty of current change. Transistors Q1–Q7 (Fig. 17.2) translate the current change to a voltage change.

"Pop" Elimination

The dial of a telephone shunts the handset during pulsing to eliminate the annoying "pop" that would otherwise occur with each dial pulse. As we have the handset "off the hook," we need to simulate this condition. The "pop stop" circuit-capacitor C4 is charged rapidly from the dial pulser and discharges only through R10. As long as the voltage on C4 is above the threshold of IC9-c, relay RY2 will be operated. The contacts of RY2 remove the telephone from the line and replace it with

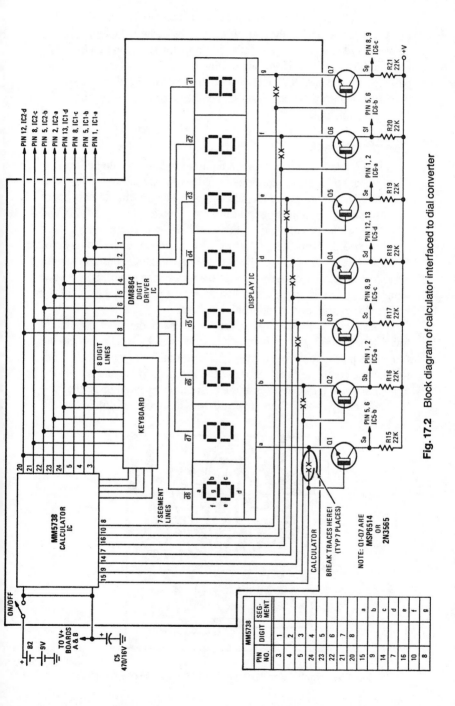

Fig. 17.2 Block diagram of calculator interfaced to dial converter

approximately 600 ohms of resistance. This makes the central office think there is a phone across the line. It also gives us an indication that the converter is dialing, since part of the 600-ohm load is composed of a LED indicator. The level at IC9-c input will slowly decay after the last dial pulse is received until RY2 finally transfers and puts the phone across the line, allowing normal conversation. The LED remains on as long as the converter is dialing.

Construction

Two printed-circuit boards are required to build the dial converter. Circuit-board patterns for the single-sided boards are shown in Figs. 17.3 and 17.4. The circuit-board method is the best approach, but perf board can also be used. Component placement is illustrated in Figs. 17.5 and 17.6. The two boards are joined with half-inch threaded spacers.

Interboard wiring is best performed with ribbon cable, but any No. 24 or 26 wire will do. Connections to the calculator are best made with very small wire, such as the No. 36 used with a wiring pencil. Solder each wire directly to the chip pads to avoid lifting the foil on the calculator board. Use a small tip on your soldering iron.

The segment lines must be broken to insert the level translating transistors Q1–Q7. Use a sharp knife, such as an X-ACTO, and cut the foil in two places about $1/16$ inch apart. Remove the $1/16$-inch piece of foil to insure a complete break. Follow the segment lines, and solder the emitter sides of Q1–Q7 wires at the read-out connection pads; the base leads to the chip pads. It's best to do all the wires group by group: first, all

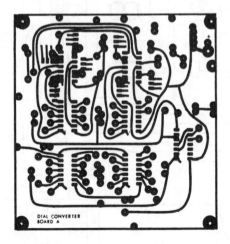

Fig. 17.3　Printed circuit pattern for board A (shown one-half size)

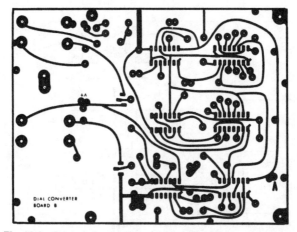

Fig. 17.4 Printed circuit pattern for board B (shown one-half size)

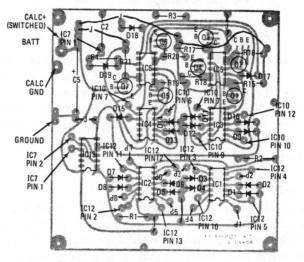

Fig. 17.5 Component layout for board A

the digit leads; then, all the segment-emitter leads; etc.; connecting each at both ends before proceeding to the next wire.

The converter can be housed in a cutdown LMB type 145 chassis box, but almost anything that will contain the circuit boards and calculator will do. The LMB 145 box is 7 inches long, 5 inches wide, and 3 inches high. Its height can be cut down to ¾ inch, and 7½ × 2⅛ × 3/16-inch wood strips can be painted black and bolted to the sides.

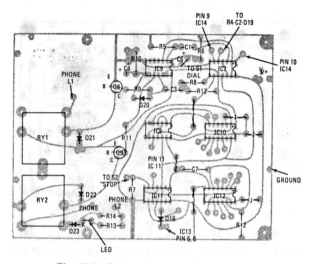

Fig. 17.6   Component layout for board B

Calculator Selection

The calculator used in the model was a surplus Bowmar purchased as a kit. Many other calculators that use the MM5738 now cost about the same as the kit. The Novus 850, 823, and 826 are good examples of a low-cost eight-digit display with memory that uses the MM5738. You can, of course, buy a calculator IC, driver, read-out, and keyboard and lay out your own calculator circuit board, but it's a lot easier (and probably cheaper) to let someone else do it for you.

How to Use the Converter

Connect the converter to the telephone line as shown in Fig. 17.7. (Reverse the L1 and L2 or the red and green connections if the LED does not light when dialing.) Turn on the calculator and clear it. Enter the telephone number you wish to call just as you would normally dial it. The read-out verifies the number you have entered. Lift the telephone receiver and press the dial button after you hear the dial tone. The selected number will quickly be dialed. If the line is busy, either place the number in memory and clear the display (to conserve the battery) or leave the number in the display. To redial, just press the dial button (or recall the number and press the dial button). A direct-dialed long-distance number must be handled in two steps. Enter the telephone number and store it in memory. Clear the display and enter the access and area codes. Press the dial button and wait until the access and area codes have been dialed; then

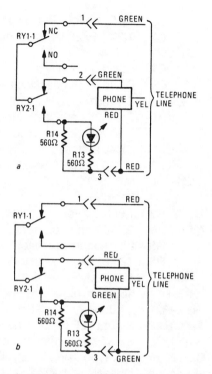

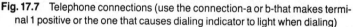

Fig. 17.7 Telephone connections (use the connection-a or b-that makes terminal 1 positive or the one that causes dialing indicator to light when dialing)

recall the telephone number and press the dial button again. To redial, all you have to re-enter is the access and area codes. The telephone number remains in the memory.

The calculator can be used normally anytime it is not dialing.

NOTE: The MM 5738 is constructed to blank the display approximately 16 seconds after the last key activity has occurred. This feature can be disabled by tieing pin 1 to V_{DD}. (It may be desirable to do so to avoid the sudden and embarrassing loss of the number you are dialing right in the middle of a digit.)

Parts List

C1, C3	0.1-μF, 25-volt ceramic capacitors
C2	.01-μF, 25-volt ceramic capacitor
C4	1-μF, 35-volt electrolytic capacitor
C5	470-μF, 16-volt electrolytic capacitor

C6, C8	1-μF, 16-volt electrolytic capacitor
C7	.022-μF, 25-volt ceramic capacitor
D1–D22	1N914
IC1, IC2, IC7	CD4011 quad two-input NAND gate
IC3, IC4	CD4030 quad exclusive-OR gate
IC5, IC6, IC8, IC9, IC13	CD4001 quad two-input NOR gate
IC10	CD4026 decade counter decoded seven-segment outputs
IC11	CD4013 dual type-D flip-flop
IC12	CD4015 dual four-bit shift register
Q1–Q8	MPS6514, 2N3565
Q9	2N3638
R1, R2, R3	1,000,000-ohm resistors*
R4, R6, R7, R9, R11	10,000-ohm resistors*
R5	2.2-megohm resistor*
R8	220,000-ohm resistor*
R10, R12	470,000-ohm resistor*
R13, R14	560-ohm resistor*
R15–R21	22,000-ohm resistor*
RY1, RY2	Sigma 65F (9 volts dc, PC mount)
S1, S2	SPST momentary pushbutton switches
Calculator	Any type using MM5738 calculator IC
Miscellaneous	Chassis, ribbon cable, spacers, LED indicator

*All resistors are ½-watt, 10% unless noted.

18

Pushbutton Dialer with Memory

By using three MOS integrated circuits from General Instrument and a demonstrator PC board layout, the construction of a deluxe telephone dialer becomes a fairly routine procedure. The circuit is a pulse dialer that interfaces with a telephone-type 2-of-7 keyboard, or, with the addition of a diode encoder, a 1-of-12 calculator keyboard.

The dialer has three basic modes of operation. First, it converts any conventional dial phone into a pushbutton phone, in which a series of up to 20 digits are stored and sent out sequentially at a fixed pulse rate.

Second is the very convenient redial mode. If the number you dial is busy, you hit the redial key twice to redial the number automatically, that is, without having to re-enter the digits. The first push of the redial key holds the last number dialed in a series of memory registers while the hook switch is depressed to get another dial tone. The second redial key closure starts the actual dialing.

Third, the system has storage for ten 20-digit numbers, including access pauses. Access pauses are required when dialing code prefixes are used to connect through automatic telephone routing systems. Often you must wait for dial tones after these codes are entered. The dialer stops the dialing sequence when it reads an access pause code from memory. Upon receipt of the next dial tone, the Continue button is pushed to finish the dialing sequence or to resume the dialing until the next access pause code.

Before getting too deep into this project, accept a word of caution. If you add this gadget to a privately owned home or company in-house phone system, you're on firm ground. But be reminded that the telephone company tends to be a little fussy about things being hooked into their lines. This device is not intended to be connected directly to a telephone set without the subscriber's compliance with local phone company regulations.

How It Works

Figure 18.1 shows the schematic diagram of the telephone dialer. Pushbutton to dial-pulse converter IC4 is the focal point of the system. A logic zero on the reset input (pin 3) clears all internal shift register stages

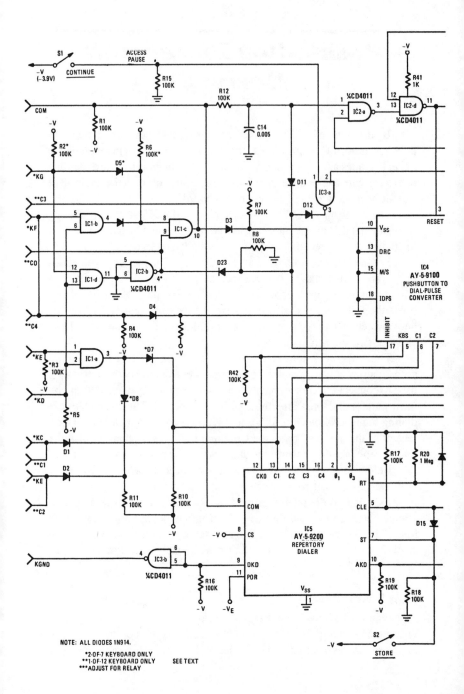

NOTE: ALL DIODES 1N914.

*2-OF-7 KEYBOARD ONLY
**1-OF-12 KEYBOARD ONLY   SEE TEXT
***ADJUST FOR RELAY

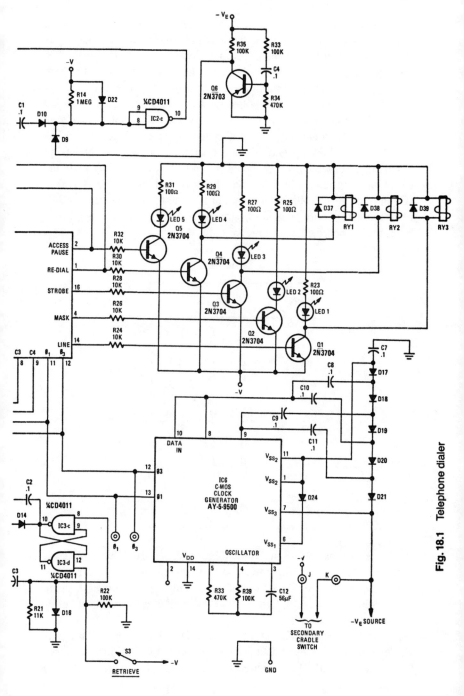

Fig. 18.1 Telephone dialer

and resets the counters. Transistor Q6 is turned on for a short interval when V_E is switched on by the hook switch. Base current to Q6 flows through R33 and C4 for the time it takes capacitor C4 to charge. The collector of Q6 remains low for the short interval and then switches high to trigger the monostable multivibrator formed by IC2-c and IC2-d.

The pushbutton to dial-pulse converter IC4 accepts a keyboard parallel input on lines C0 through C4, coded as listed in Table 18.1. Figure 18.2 shows the connections for a telephone-type 2-of-7 keyboard. The C1, C2, C3, and C4 inputs of IC4 are all negative or logic-1 levels except when pulled down by the keyboard outputs. Hitting any key pulls the COM line to ground, which activates the keyboard strobe input (KBS) of IC4 through the COM input terminal of IC5. Ten milliseconds later, IC4 reads the state of parallel inputs C0 through C4. This debouncing interval gives the keyboard contacts time to settle.

Table 18.1

Digit	C1	C2	C3	C4
1	1	1	1	1
2	1	1	1	0
3	1	1	0	1
4	1	0	1	1
5	1	0	1	0
6	1	0	0	1
7	0	1	1	1
8	0	1	1	0
9	0	1	0	1
0	1	1	0	0
Access pause	0	0	1	1

Pressing key "1" grounds only KE, which is *one* of the two inputs to AND gate IC1-a. The output of IC1-a goes to logic 0 only when both its inputs are at a logic 0. Since pin 2 of IC1-a is high, the output of this gate remains high. Therefore, C1 through C4 are all at a logic 1 level corresponding to digit 1 in Table 18.1.

Key "2" brings KF and C4 to ground. Depressing key "3" grounds C3 through IC1-c. Keys "4" through "9" work by their direct connection to C1, C2, and C4 and indirect connection to C3 through IC1-c.

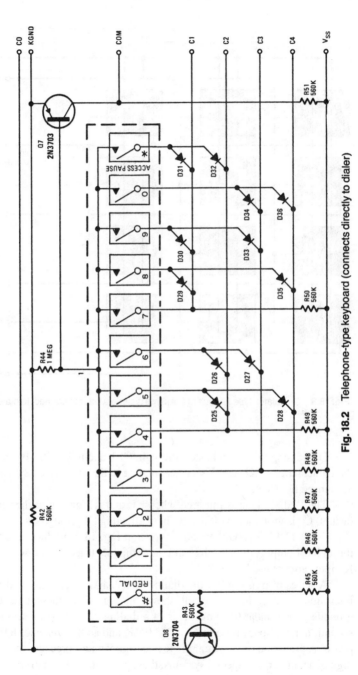

Fig. 18.2 Telephone-type keyboard (connects directly to dialer)

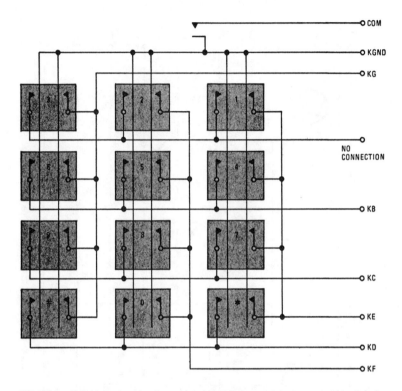

Fig. 18.3   Calculator-type keyboard (requires dialer encoder to connect to dialer)

Depressing "0" switches C3 through IC1-b and IC1-c. IC1-b senses the coincidence of KD and KF corresponding to the "0". Access pauses are sensed by IC1-a.

The Redial mode is initiated when KE and KF go low. When this occurs, C0 is grounded by IC1-d without affecting C1 through C4.

The 1-of-12 keyboard encoder shown in Fig. 18.3 produces the C0 through C4 outputs directly. IC1 and the components associated with the K inputs are not used.

A series of pins control the dialing rate, mark/space ratio, and the interdigital pause. In the circuit shown in Fig. 18.1, these pins are grounded for standard timing. This is a dialing rate of 10 pulses-per-second, a mark/space ratio of 66⅔%/33⅓%, and an 800-ms interdigital pause. A predigital pause equal to the interdigital pause precedes the first digit of a number. For special systems, these pins can be wired to either of the two clock phases or logic 1 to change the parameters.

Don't all integrated circuits have a power supply pin? Not this one! Energy is supplied to IC4 from the two clock inputs, ql and q2. The clocks must swing at least 13.5 volts negative and are produced by a special clock generator IC.

The inhibit input has the dual purpose of inhibiting the dial pulses when access pauses are required and initiating a redialing output. The remaining pins are the outputs that drive the LED indicators and output relays.

The repertory dialer, IC5, is the ten-number memory. Although it has the capability for 22-digit storage when used in touch-tone systems (using other GI ICs), the phone number length is limited to 20 digits in this circuit by IC4. Although the circuit in Fig. 18.1 uses only one AY-5-9200 IC for a total storage capacity of ten telephone numbers, this IC was designed to be "stacked" for additional storage capacity by paralleling the inputs and outputs and using the chip-select input (pin 8) to select the memory block.

The COM output of the keyboard feeds the COM input (pin 6) of IC5. The COM input is transmitted to the AY-5-9100 through its KBS input only when a dial or redial operation is in progress.

During a store operation, the keyboard signals are entered into the AY-5-9200 and the CKO line (IC5, pin 12) is inhibited so the signals on C0–C4 do not cause any dial pulses to be transmitted. The CLE line (pin 5) is activated at the same time as the store line (pin 7). The first digit then depressed is latched as the memory address, and that location is cleared. The number to be stored is entered into the location and the store button is released. The CLE line is simultaneously released.

The retrieve mode isselected by applying a logic 1 level to the CLE input and pulsing the retrieve input for at least 10 ms through capacitor C2 by the flip-flop formed by IC3-c and IC3-d. The following digit entered on the keyboard is latched as the memory address. The dial pulses are transmitted at least 60 ms later.

Address keyboard disable (AKD) output line (pin 10) is held at a logic 0 level during the store and retrieve operations. The positive-going transition at the end of a retrieve operation resets flip-flop IC3-c/IC3-d. IC5 is also powered from the two-phase clock signals. IC5 is cleared on initial turn or when $-V_E$ is applied to pin 11.

The CMOS clock generator (IC6) is wired as a voltage multiplier to convert the 3.9-volt supply to nominal 15-volt clock outputs using a Cockroft-Walton voltage multiplier. An internal D-type flip-flop is connected as a divide-by-two by tying the Q output to the D input using the jumper between pins 8 and 10. The Q output drives capacitors C8 and C10, and the Q̄ output drives capacitors C9 and C11. Diodes D17 through

D21 and capacitors C8 through C11 boost the −3.9-volt supply to −15 volts on capacitor C7 and power the clock generator on pins 1 and 11.

Construction

Figures 18.4 and 18.5 are the component- and bottom-side PC board foil patterns. Figure 18.6 is the component placement diagram. You have the option of using either one of two types of keyboards. The dialer uses the standard 0 through 9 keys plus two more for Redial and Access pause. The actual number of switches on the keyboard will exceed 12 if the calculator is designed for extra functions.

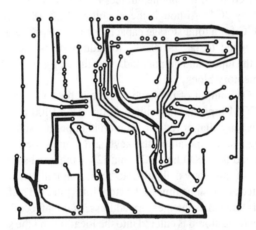

Fig. 18.4   Foil pattern of component side of PC board

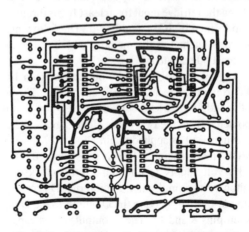

Fig. 18.5   Foil pattern of bottom side of PC board

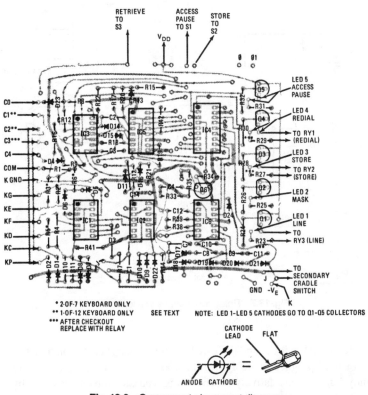

Fig. 18.6 Component placement diagram

Keyboards listed as telephone keyboards are usually the 2-of-7 type (Fig. 18.2). Conventional telephone keyboard layouts include digits 0 through 9 plus * and # keys for a total of 12 keys. Each key has dpst contacts that are switched along a matrix of three vertical busses (KG, KF, and KE) and four horizontal busses (KB, KC, KD, and the one marked "no connection"). Pressing any key makes contact with one horizontal and one vertical bus. The total of seven busses and two contacts per key accounts for the 2-of-7 designation.

All six ICs are used if the telephone-type 2-of-7 keyboard is used. If the calculator-type keyboard is used, a separate encoder is needed. In this case, IC1 is eliminated since its purpose is to encode the 2-of-7 keyboard signals. The parts list and the diagrams reflect the component variations in the two types of keyboard.

Transistors Q1 through Q5 are the output drivers that control the five LEDs and the relays. Typical phone connections are shown in Fig. 18.7.

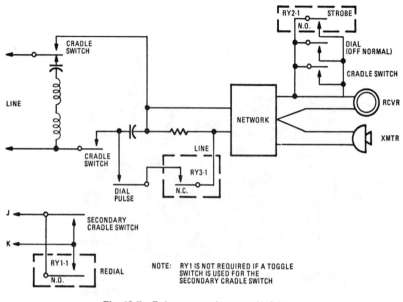

Fig. 18.7 Relay connections to telephone

Relays RY1, RY2, and RY3 are best mounted right in the phone. The author used Magnecraft relays with a 100-ohm coil because they were handy, but they are relatively expensive and you can probably do better by looking around. The supply feeding the emitters of the output transistors can be isolated and increased in voltage if you need more than 3.9 volts for your relay section. Resistors R25, R27, R29, and R31 will have to be changed accordingly.

Two normally open relay contacts — the strobe and the redial contacts — and a third normally closed line relay are needed. The relays are connected to the collectors of Q2, Q3, and Q4. After a checkout of the system, you may elect to remove the LEDs.

Either IC sockets or Molex pins should be used to mount the ICs. If you have to replace a defective IC or remove one for troubleshooting, you'll be glad when it comes out easily.

Unless you go to the trouble of making your PC board with plated-through holes, you will have to solder the components to the foil on both sides of the board. Jumpers must be inserted and soldered to both sides in all empty holes that connect to foil runs.

IC2 is a CMOS device. When it is not used, the input of IC2-b becomes unterminated and must be grounded for the proper operation of the other gates in the IC2 package. A short jumper is added on the rear of the board. The output of the gate must also be disconnected so that it does not interfere with the C0 keyboard output. The best way to do this is simply to leave out the jumper between the front and rear of the boards (indicated with an asterisk in the component placement diagram).

The telephone dialer described here uses a 1-of-12 type keyboard with the encoder mounted on a Veroboard. The parallel conductor runs of the Veroboard are perfect for matrix circuits like those of the encoder shown in Fig. 18.3. A specific layout for the encoder board has not been presented since the layout depends on the particular pin arrangement of the keyboard. Again Molex pins are recommended so that the keyboard can be mounted right over the encoder components, yet can be easily removed for troubleshooting.

Momentary pushbutton switches are used for the Store, Retrieve, and Continue functions. An additional hook switch contact is needed to apply power to the clock generator, IC6. If a spare normally closed contact is not available on the cradle switch, a microswitch can be rigged to the bracket switch assembly. Although somewhat less convenient, a separate toggle switch can be used. Relay RY1 is not required if a toggle switch replaces the secondary hook switch.

The system is powered from a 3.9-volt (± 5 percent) negative voltage supply. A zener-regulated supply will do the job. Remember that the number memory is volatile and that power must be kept on continuously. The supply should not be designed to supply more than 200-mA peak current drain of the LEDs and relays. Standby power drain is very low, essentially only the 2.25-mW drain typical of IC5.

Checkout

Once everything is together, you will be anxious to put the circuit through its paces. Initial testing consists of watching the response of the LEDs to pushbutton sequences.

Connect the -3.9-volt supply to the $-V_E$ pad on the PC board, and the power-supply ground to the GND pad. Turn on the power and flip the hook switch. Operating the switch simulates lifting the receiver and resets the registers in IC4.

Now try keying in a number. Each key closure stores the corresponding digit in internal registers and dials them out with precise timing. Because of the memory, the keys can be pressed at a faster rate than they are dialed out.

The mask, strobe, and line LEDs should operate in the following way: The mask LED should be lit during the entire dialing sequence. The strobe LED is illuminated during the time it takes to dial out the series of pulses that make up one digit. The line LED will flash once for each output pulse; each key pressed will flash this LED as many times as correspond to the numeral printed on the key (zero flashes the LED ten times). Of course, the flash rate is 10 Hz so that you will not be able to count the individual pulses by eye, but you can roughly discern the shorter and longer sequences.

After verifying the individual digit operation, check the redial facility. On the calculator-type keyboard used here, the C (constant) button was wired and used as the redial key. Dial a sequence of digits representing a phone number, and then press the redial key. The redial lamp should light in preparation for sending the number. Hit the redial key again to start the redial pulsing. Redialing can be repeated as many times as desired. The reason for the flip-flop action is that in normal use, the hook switch must be cycled before redialing to get a new dial tone. Since the hook switch is in series with the power supply to IC4, it interrupts the power and resets the IC4 registers. The first operation of the redial key closes a relay contact that parallels the hook switch. At this point, the receiver can be hung up and picked up again without interrupting the power to, and resetting, IC4. Now when the redial button is pressed a second time, the power-bypass relay contact across the hook switch is disconnected, and the transmission of the stored digits begins.

Proceed to the checkout of the ten-number memory. Numbers are stored by pressing the store button and continuing to press it for the entire store operation. The first digit entered is the storage address. It can be any digit from 0 to 9, each of which represents of the ten storage locations in memory. The digits that follow the address are the numbers to be stored. The keyboard used here was wired so that the access pause key would serve as the decimal point. Be sure to intermix one or two of these pauses with the test numbers.

After the first number is entered, flip the hook switch back and forth, but fairly slowly, to give the reset capacitor time to charge. Then press the store button, the next storage location, and the next phone number. Repeat this sequence to a total of ten times for the numbers you want to store. Switch the temporary-hook switch of the off-hook position, and get ready to recall your first number by pressing the retrieve button. In contrast to the store button, the retrieve switch is closed momentarily and does not have to be held. The next digit entered addresses the memory originally tagged by the same digit during the store operation and begins the dialing of the stored number.

When an access pause is reached, the system stops, with the access pause LED lit. Momentary closure of the continue switch should resume the dialing sequence.

If everything checks out at this point, you are ready to complete the relay connections to your phone as shown in Fig. 18.7. Retain the redial and access pause LEDs in your system as visual aids for these functions.

Parts List

C1–C4, C7–C11	0.1-μF, 50-volt disk
C12	56-pF, 50-volt disk
C14	.005-μF 50-volt disk
D1–D4, D5–D8*, D9–D24, D25**–D36**, D37–D39	1N914
IC1*	CD4081, quad two-input AND gate
IC2, IC3	CD4011, quad two-input NAND gate
IC4	AY-5-9100 (General Instrument)
IC5	AY-5-9200 (General Instrument)
IC6	AY-5-9500 (General Instrument)
LED1–LED5	MV5053 (Monsanto)
Q1–Q5, Q8	2N3704
Q6, Q7*	2N3703
R1, R2*, R3*, R4, R5*, R6*, R7-R12, R15–R19, R22, R33, R35, R39, R42	100,000-ohm resistors
R14, R20, R21, R44**	1-megohm resistors
R23, R25, R27, R29, R31	100-ohm resistors
R24, R26, R28, R30, R32	10,000-ohm resistors
R34, R38	470,000-ohm resistors
R41	1000-ohm resistor
R42**, R43**, R45**–R51**	560,000-ohm resistors
RY1, RY2	spst normally open relay, 100-ohm coil (Magnecraft 103MX-10 or equal)
RY3	spst normally closed relay, 100-ohm coil (Magnecraft 103MX-10 or equal)
S1-S3	spst normally open

Note: The following component designations are not used and do not appear in the parts list, layout, and schematic: R13, R36, R37, R40, C5, C6, and C13. All resistors are ¼-watt, 10 percent, unless noted. For explanation of asterisks, see Fig. 18.1.

19

Alarm Dialer

Protection of personal property is an ever growing concern, with more people than ever installing burglar alarms in their houses. Most of these alarms are of the local type; if an illegal entry is made, a bell on the premises rings, and hopefully the noise will scare the burglar away. Unfortunately, the burglar may just disable the bell instead. There are other times you might like to be alerted to a dangerous condition where you are not present. For example, if you have a house in the mountains subject to winter freeze-ups and cracked pipes, you'd certainly like to know if your oil burner has shut off for some reason. Wouldn't it be nice if the burglar alarm or the shut-off oil burner could call you up to let you know what is going on? If you are willing to spend between $100 and $200, you can get a special telephone dialer device that will do just that. But for less than $20 you can build an alarm dialer circuit that uses any inexpensive cassette tape recorder to do the dialing.

As the various telephone companies strive to bring better service to their customers, they are rapidly changing over central office equipment to electronic switching systems. Among the advantages of this new equipment is that it makes Touch-Tone dialing possible. In areas that have converted to electronic switching, and they are rapidly expanding, a customer may pay an extra monthly fee to get a tone-operated phone. Many people don't realize that Touch-Tone dialing is already available to them for the asking. The dialer discussed in this project requires that you have Touch-Tone service.

Theory of Operation

In Touch-Tone dialing, pairs of audio frequencies are used to signal the dialing of a particular number. In the more conventional rotary dialing system, dialing the number 6, say, results in six pulses being generated and sent to the central office for interpretation. In Touch-Tone dialing, a unique pair of audio frequencies is generated for each digit. The frequencies generated range from slightly less than 700 Hz to a little more than 1,500 Hz. A total of seven different frequency tones are used, four low and three high, as follows: 697, 770, 852, and 941 Hz (low), and 1209, 1336, and 1477 Hz (high). The tones required for each digit are shown in Table 19.1.

Table 19.1

| Low tone group | High tone group | | |
	1209 Hz	*1336 Hz*	*1477 Hz*
697 Hz	1	2	3
770 Hz	4	5	6
852 Hz	7	8	9
941 Hz	*	0	#

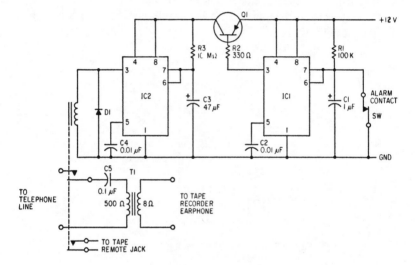

Fig. 19.1   Alarm dialer controller

The telephone company took great pains when designing the system to keep the tones in the audio range so that existing telephone circuits could be used. The system works to our advantage because these signals are easy to record and play back even on inexpensive tape recorders.

When using this dialer, you must first record the tones generated by the telephone for the particular number that you want dialed. Then the audio message desired is recorded a few times. If you add a little more circuitry to the unit, it becomes possible to hang up and dial other numbers to relay the same or a different message.

The key to this dialer is the dialer controller shown in Fig. 19.1. This circuit uses two inexpensive 555 IC timer (or one 556 dual timer integrated circuits to detect an alarm condition, turn on the recorder that performs the dialing and relays the message, and then turn the recorder off.

Both timer ICs are connected in the monostable multivibrator mode, using a special configuration called a *power-up monostable* (see Fig. 19.2). The power-up monostable is so called because it prevents a pulse from appearing at the output (pin 3) of the timer until a certain amount of time ($t = 1.1RC$) has passed after power is applied. When power is applied, the capacitor starts charging through the resistor. The junction of these two components is connected to the junction of pins 6 and 7 of the 555, which are the threshold and discharge pins. When the voltage applied to pin 6 rises to a value that is greater than two-thirds of the value of the voltage applied to pin 8, a flip-flop inside the timer IC is reset, and the output of the timer goes low.

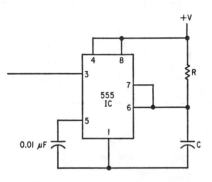

Fig. 19.2   Power-up monostable multivibrator

In the dialer controller, the first timer circuit (IC1) is used to detect the alarm condition. It is configured as a power-up monostable that waits only about 0.1 sec before driving output pin 3 from the high to the low state. The alarm contact is placed in parallel with C1 and should normally be closed, thereby shorting out the capacitor and insuring that a positive voltage is applied to the base of the transistor (Q1), keeping it in the off state.

When an alarm condition exists, this contact must be opened. When the short is removed, the capacitor charges up to 8 volts and triggers the timer, driving the output low and turning on transistor Q1. When Q1 conducts, it applies power to IC2, another power-up monostable. This monostable's output is initially high, and power is applied to the relay that controls the cassette recorder, turning it on. The output of IC2 stays high for about 8.5 minutes ($t = 1.1$ R3C3), which is enough time to dial the number desired and give a reasonably long message. The time the recorder stays on can be varied by changing either R3 or C3 or both. After 8.5 minutes, the output of IC2 will go low (because the voltage across C3 has reached 8 volts), thereby shutting off the relay, which shuts off the tape recorder in turn and hangs up the phone.

Interfacing to the Telephone

Probably the simplest part of this project is interfacing the unit to the telephone. To do so, merely take the two wires marked "To Phone Line" and connect them in parallel with the red and green wires coming from your telephone. You can make the job a little easier, and also make your dialer mobile, by connecting these wires to a plug/jack connector of the sort sold in most telephone and hobby electronic stores. You don't have to worry about your telephone at all, because the relay that turns the tape recorder on and off also connects and disconnects the telephone. When the relay closes, it connects a series circuit composed of C5 and T1 (the 500-ohm winding). This circuit looks like an off-hook telephone and thus terminates the line and permits dialing.

Checking It Out

To check out the dialer, you must first record the dialing tones from a Touch-Tone telephone. To do so easily, attach a telephone pickup coil to the handset at the receiver end. Now lift the receiver off its cradle and start your tape recorder in the Record mode, making sure that you are past the nonmagnetic leader on the cassette. With the tape recording, punch in the digits of the number you want dialed; then immediately hang up before the call has a chance to be completed.

Now that the number to be dialed has been recorded, you can record the message you want sent. It is advisable to repeat the message a few times so that if the phone is not answered right away, the called party can still hear the full message. In accordance with the length of the message, you can change the amount of time available by adjusting R3 and C3.

Once you have finished recording your message, you can test the dialer out. The only possible problem you will have is that the phone may not be dialed because the volume is not right. Plug the dialer in and short out C1. With the volume on your tape recorder adjusted to about mid range, the tones should be sufficiently loud to cause the telephone to be dialed. If it isn't, just increase the volume. After the number has been dialed, the message should be given and the dialer should hang up when the time determined by R3 and C3 has elapsed.

Parts List

C1	1-μF electrolytic capacitor (16 WVDC)
C2	0.01-μF capacitor
C3	47-μF electrolytic capacitor (16 WVDC)
C4	0.01-μF capacitor
C5	0.1-μF capacitor
D1	1N4148
IC1, IC2	555 timer IC
R1	100-kilohm resistor
R2	330-ohm resistor
R3	10-megohm resistor
RY1	12-volt dpst relay
SW1	normally closed alarm contact in active position
T1	500- to 80-ohm audio output transformer

20

Touch-Tone™ Encoder*

For those of you who want the advantages of pushbutton dialing without the extra cost of a special phone, this little circuit may be the answer. It is a Touch-Tone^{tm} encoder built with the help of the ubiquitous 555 timer IC. In actuality, two 555 timers are needed; if desirable, a 556 dual timer can be used to reduce package count.

Theory of Operation

In Touch-Tone dialing, each pushbutton pressed generates a pair of frequencies in the audible range from just less than 700 Hz to slightly less than 1500 Hz. The frequency pairs consist of two of seven possible tones. As was seen in Table 19.1, these tones are divided into a "low" group (rows) — consisting of frequencies 697, 770, 852, and 941 Hz — and a "high" group — consisting of frequencies (columns) 1209, 1336, and 1477 Hz. These are the frequencies used in the standard 12-button Touch-Tone telephone. Sixteen-button sets have an additional column tone.

To generate these two groups of tones, two oscillators built from 555 timer ICs are required. Each is connected in the astable configuration and produces a frequency determined by the following formula:

$$f = 1/[0.693(R_a + 2R_b)C]$$

where R_a is the resistance between the timer discharge output (pin 7) and the positive side of the power supply (V_{CC}) and R_b is the resistance between the threshold input (pin 6) and the discharge output. As shown in Fig. 20.1, R_a is replaced by resistive divider string for both the low- and the high-tone oscillators.

IC1 is the low-tone oscillator. To calculate the resistor needed for the 941-Hz tone, let C = 0.047 μF and R_a = R1 = 4.3 kilohms. Solving the equation for R_b, we find it be 14,164 ohms. To generate the next lower tone (852 Hz), R_a = R1 + R2 so that R2 = 3.3 kilohms. For the 770-Hz tone, R_a = R1 + R2 + R3, so that R3 = 3.9 kilohms. In a like manner, R4 = 4.3 kilohms.

*Adapted and reprinted with permission from the August 1977 issue of Ham Radio Magazine, © 1977 by Communications Technology, Greenville, New Hampshire 03048.

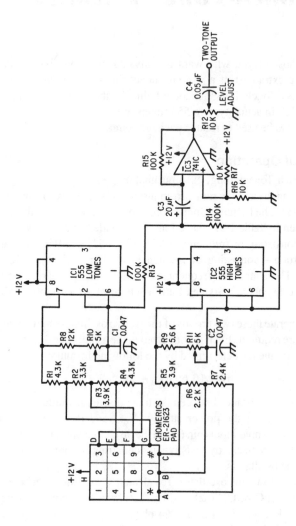

Fig. 20.1   Touch-Tone encoder schematic

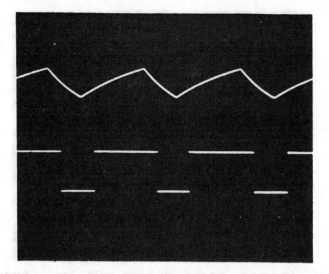

Fig. 20.2 Type 555 timer output waveforms (when connected as an astable multivibrator, upper trace is obtained from pins 2 and 6; lower trace, from pin 3)

The high-tone oscillator IC2 is designed in a similar manner. Starting with the 1477-Hz tone, if we let $R_a = R5 = 3.9$ kilohms and $C = 0.047 \mu F$, then R6 = 2.2 kilohms and R7 = 2.4 kilohms.

For both oscillators, the outputs are taken from the timer discharge junction and trigger pins (pins 2 and 6), which produce a pseudotriangular waveform between ⅓ and ⅔ Vcc (Fig. 20.2). A 741C op-amp, IC3, adds the output of both oscillators, shown in Fig. 20.3, and is coupled to a 10-k potentiometer used as an output level control. From here the signal can be fed either to an audio amplifier and speaker, where it can be audibly coupled to the telephone, or it can be transformer-coupled directly to the telephone line.

Components

For good thermal stability, the 0.047-μF capacitors should be either tantalum or mylar, and resistors R1 through R9 should be rated at 1 percent. Several manufacturers currently advertise a 4-by-3 key pad similar to the Chomerics type ER-21623. Potentiometers R10 and R11 are 10-turn trim pots. If desired, a 556 dual timer can replace ICs 1 and 2. Figure 20.4 compares the pin connections of the 555 and the 556 timers.

Adjustment

Start the adjustment procedure by pressing the * key and adjusting R10 so that the low-group oscillator is set at exactly 941 Hz. You'll have to

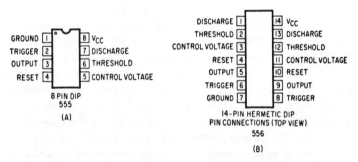

Fig. 20.3 Addition of low-tone oscillator (top trace) and high-tone oscillator (center trace) gives two-tone output signal (bottom trace) when the digit 2 is pressed

GROUND 1 — 8 V_{CC}
TRIGGER 2 — 7 DISCHARGE
OUTPUT 3 — 6 THRESHOLD
RESET 4 — 5 CONTROL VOLTAGE

8 PIN DIP
555
(A)

DISCHARGE 1 — 14 V_{CC}
THRESHOLD 2 — 13 DISCHARGE
CONTROL VOLTAGE 3 — 12 THRESHOLD
RESET 4 — 11 CONTROL VOLTAGE
OUTPUT 5 — 10 RESET
TRIGGER 6 — 9 OUTPUT
GROUND 7 — 8 TRIGGER

14-PIN HERMETIC DIP
PIN CONNECTIONS (TOP VIEW)
556
(B)

Fig. 20.4 Pin-connection comparison for type 555 and 556 timers

make this adjustment with the aid of either an oscilloscope or a frequency meter. (The latter is preferred.) The measurement for the low-group oscillator is made at pin 3 of IC 1. After this adjustment has been made, the frequencies of 852, 770, and 697 Hz should be obtained to within 2 percent when numbers 7, 4, and 1 are pressed, respectively. If you don't get these values, then the value of resistors R2 through R4 are off.

For the high-tone group, press the # key and adjust R11 so that the oscillator is set at 1477 Hz on pin 3 of IC 2. Consequently, frequencies of 1336 and 1209 Hz should be obtained within 2 percent when the 0 and * keys are pressed. A comparison of the digit 1 generated by a telephone company Model 35 pad and the 555 circuit described here is shown in Fig. 20.5.

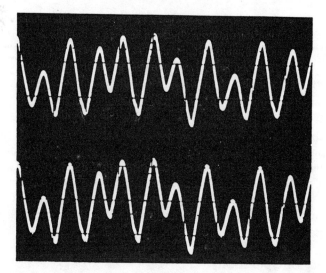

Fig. 20.5 Comparison of two-tone output obtained from 555 timer encoder (upper trace) with Western Electric Model 35 pad (lower trace) for the digit 1

Parts List

C1, C2	0.047-μF capacitors
C3	20-μF capacitor
C4	0.05-μF capacitor
IC1 – IC2	555 timer
IC3	741 operational amplifier
R1, R4	4.3-KΩ resistors
R2	3.3-KΩ resistor
R3, R5	3.9-KΩ resistors
R6	2.2-kΩ resistor
R7	2.4-kΩ resistor
R8	12-kΩ resistor
R9	5.6-kΩ resistor
R10, R11	5-kΩ resistor
R12	10-kΩ potentiometer
R13 – R15	100-kΩ resistor
R16, R17	10-kΩ resistor
Miscellaneous	CHOMERICS ER-21623 (or equivalent) key pad

21

Phone Meter

One way of keeping tabs on a telephone line is to monitor the voltage across it. The normal voltage across an unterminated line is 48 volts. If any bugging devices are added to the line, there is a good chance that they will alter the voltage on it by drawing some current for their own operation. In addition, with this phone meter, you can also see right away if anyone has picked up an extension phone and is eavesdropping on your conversation.

Theory of Operation

The circuit for the phone meter (Fig. 21.1) is very simple and consists of only four components: a 50-μA meter, a diode bridge, a switch, and a 1-megohm resistor. Operation is simple. The series resistor

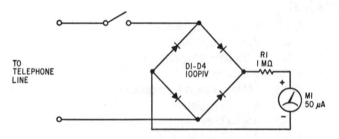

Fig. 21.1 Phone meter circuit

is used to calibrate the meter so that each microamp represents 1 volt. Thus, for full-scale deflection, 50 volts are required, a figure determined by Ohm's Law as follows:

$$E = IR$$
$$= (50 \times 10^{-6} \text{ amps}) (1 \times 10^{6} \text{ ohms})$$
$$= 50 \text{ volts}$$

You can change the full-scale deflection of the meter simply by changing the series resistance. For example, a 2-megohm resistor will convert the meter to a 100-volt, full-scale deflection meter.

Construction and Checkout

With only three components, this is the simplest project in this book to build. And success is virtually guaranteed. Construction time should be no more than five minutes. When putting the meter together, it is suggested that a full-wave bridge rectifier module be used instead of the individual diodes to reduce the component count and simplify matters. The positive terminal of the bridge gets connected to the resistor, which in turn is connected to the positive side of the microammeter. The negative side of the bridge goes to the other side of the microammeter. The ac inputs of the bridge rectifier go to the telephone line. Because of the bridge, they are not polarity sensitive; therefore, either lead can be connected to the green wire and the other to the red.

Once put together and housed in a suitable enclosure, the meter can be checked out by connecting it to the telephone line. After closing the switch, the meter should read about 48 volts (varying a volt or so, depending on your particular system). Now lift the receiver off the hook. The voltage should drop to somewhere around 5 volts, depending on your telephone instrument. Now if an additional phone is picked up, the voltage will drop even more, making it easy for you to tell if someone has picked up an extension. You should check the voltage daily. If there are any significant changes, chances are that a bug has been placed on your line.

Parts List

D1–D4 silion diodes 100 PIV or higher
M1 0- to 50-μA meter
R1 1-megohm, ½-watt resistor

22

Antibug Monitor

In this age of advanced electronics, it is a relatively simple matter to buy or build a device that is capable of bugging a telephone. Needless to say, it is often desirable to use a phone secure in the knowledge that no bugging devices are attached. Now, for an investment of less than $15 and only half an hour's labor, you can have that secure feeling. This antibug monitor, once connected to a known clear line, will inform you instantly if someone cuts the line in order to install a series bugging device. It will not, however, detect parallel bugs. Other techniques described elsewhere in the book will allow you to do that.

Theory of Operation

The heart of the antibug monitor is a silicon-controlled rectifier, which, once triggered, stays on until it is reset. An SCR is turned on by momentarily applying a positive voltage to its gate. It can be turned off in a number of ways. One is to interrupt the lead going to the anode or the cathode of the SCR momentarily. Another is to place a temporary short between the anode and cathode. A third is to make the anode more negative than the cathode. It is the latter approach that will be helpful here.

A LED and a current-limiting resistor are placed in series with the anode of the SCR and the positive supply voltage so that when the SCR is triggered, the LED will light up. Next, a voltage divider is connected to the gate of the SCR so that the voltage applied to the gate can be precisely controlled. The voltage divider is composed of a fixed resistor (R6) and a variable resistor (R4). In use, the variable resistor is adjusted so that the LED just barely lights up.

Connected in parallel with the voltage divider is a series RC circuit composed of R3 and C1. It is important to note that C1, if polarized, must have its negative side connected to the C1–R3 junction. Also connected to this junction is the wiper of potentiometer R2, which, along with R1, forms another voltage divider.

Circuit operation is simple (refer to Fig. 22.1). R1 is connected to the red, or negative, line of the common two-wire telephone pair, whereas R2 is connected to the green, or positive, side. Once R4 has been adjusted

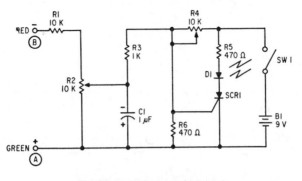

Fig. 22.1 Antibug monitor circuit

to turn the LED on, R2 is adjusted to turn it off by making the amount of negative voltage on C1 compensate for the positive voltage applied to the anode. Remember, if the voltage at the anode is less than or equal to the voltage on the cathode, the SCR will turn off.

Construction and Check out

The fabrication of this circuit is not critical, but there are a few points to which attention should be paid. The first is to make sure that the LED is connected properly, or it will not light up. It lights only when a positive voltage is applied to its anode. The second is the polarity of capacitor C1 if an electrolytic is used. If a nonpolar capacitor is used, the direction in which it is hooked up is not critical.

The entire unit can be housed in a small metal or bakelite box, and a key switch for SW1 is recommended to prevent tampering. Remember that the unit is useless if it can be easily turned off. It is also a good idea to connect it to a phone line in a fairly secure location so that it cannot be easily disconnected. A better approach would be to build it right into the phone.

To use the unit, close SW1 and adjust R4 so that the LED just barely turns on. Then atach the unit to the telephone line that is to be protected. For maximum security, a separate device is needed for each extension (although for truly maximum security, there shouldn't be any extensions in the first place). Once connected to the phone line, potentiometer R2 should be adjusted so that the LED goes out. Now, if anyone cuts the line to insert a series bugging device, the LED will come on and stay on until SW1 is momentarily shut off.

To eliminate the polarity sensitivity on the wires that go to the telephone line, the bridge rectifier shown if Fig. 22.2 can be used. In this

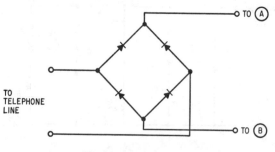

Fig. 22.2   Bridge rectifier

case, connect the positive bridge terminal to the wire labeled green on the antibug monitor and the negative bridge terminal to the wire labeled red. The other two leads of the bridge circuit can now be connected to the line in any way, there being no need to worry about polarity.

To test the circuit after it has been connected, simply disconnect one of the wires going to the unit and then reconnect it. The LED should light. Remember that the antibug monitor must be located at the farthest end of the telephone connecting wire and as close as possible to the phone in question for it to be the most effective.

Parts List

B1	9-volt battery
C1	1-μF capacitor
D1	light-emitting diode (LED)
R1	10-kilohm resistor
R2, R4	10-kilohm trimpots
R3	1-kilohm resistor
R5, R6	470-ohm resistors
SCR1	25-PIV silicon-controlled rectifier
SW1	spst key switch

23

Bug Detector

As mentioned earlier, bugging is becoming an increasingly popular sport. The previous project showed you how to build a device that would monitor a telephone line to warn you of the presence of a series bug. Although this is a commonly used bug, there are other more sophisticated ways of listening to phone conversations or eavesdropping on other tête-à-têtes. One of the most famous of these, widely used, is the *infinity bug*. Its name derives from the fact that, unlike most other devices, its range is unlimited. The reason is simple; it is connected to the telephone. When a person is interested in listening to what is happening at the bugged location, he simply dials the number and immediately activates the bug by feeding a tone from a handheld generator into the telephone. This tone automatically causes the phone to answer, even though the handset is still on the hook. Since it also disables the bell, and prevents it from ringing, no one in the room is warned that his or her conversation is being transmitted next door or around the world. This project is designed to put the infinity bug out of business.

Theory of Operation

Since an infinity bug is operated by specific tones, it is a relatively simple matter to build a device that will generate an audio signal to sweep through the entire audio range to detect it. When this signal is applied to the suspected phone line, the telephone will answer itself if a bug is present.

The audio sweep generator — a standard piece of laboratory equipment — can be purchased for about $100 or built for less than $15 like the one illustrated in Fig. 23.1. The heart of the sweep generator consists of two 555 IC timer devices (or one dual 556). Both timers are connected to operate in the astable (oscillating) mode. IC1 is designed to operate at a frequency of about 0.8 Hz, whereas IC2 is designed to operate at a maximum frequency of about 10 kHz. Frequency is determined by the following formula:

$$f = 1.44/((R1 + 2R2)C)$$

where f is in hertz, R is in megohms, and C is in microfarads.

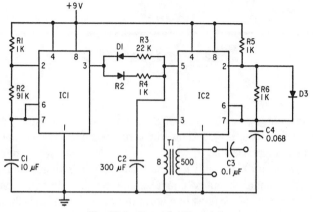

Fig. 23.1 Bug detector circuit

An interesting feature of the 555 timer IC is that it is possible to vary the frequency of oscillations by varying the voltage applied to pin 5, the control voltage input. It is this feature that allows us to produce a sweep oscillator. As the voltage on pin 5 increases, so does the output frequency. The opposite is true as well; as the voltage goes down, so does the frequency.

Normally, the output of the 555 at pin 3 is a square wave. In order to get a varying voltage on pin 5 of the second 555, we really need a triangular wave. To obtain one, we integrate the square wave produced at pin 3 of the first 555 (IC1). This integration is achieved by D1, D2, R3, R4, and C2, the latter being alternately charged and discharged.

The varying output of IC2 is then fed into the telephone line by tranformer T1, which has an 8-ohm winding going to pin 3 of IC2 and a 500-ohm winding in series with a 0.1-microfarad capacitor connected to the telephone line.

Construction and Checkout

Construction of this project is simple and straightforward. Because ICs are used, breadboard or vector board is recommended. There are no particular problems with wiring or layout since this circuit operates at audio frequencies. It is suggested that the leads being connected to the phone line (T1 and C3) end in a jack or modular connector to make the hook-up simple.

To test the device, it is necessary first to hook up the generator to a phone line (not the one suspected of containing the bug) and then dial the number of the suspected line. Instructions should be given to a party at the other end not to lift up the handset to answer the phone. Now the unit should be turned on. Approximately three times every four seconds, the oscillator will sweep up to 10 kHz and back down again. If a frequency sensitive bug is on the line, it will be activated, and the telephone will stop ringing without anyone's picking it up. You should now be able to hear everything said in the room. If the phone doesn't answer, chances are pretty good that no infinity bug is connected unless the bug requires a mulitfrequency trigger signal. These types of bugs, however, are very rare.

Parts List

C1	10-μF electrolytic capacitor (25 WVDC)
C2	300-μF electrolytic capacitor (25 WVDC)
C3	0.1-μF capacitor
C4	0.068-μF capacitor
D1–D3	1N914
IC1, IC2	555 timers
R1, R4–R6	1-kilohm resistors
R2	91-kilohm resistor
R3	22-kilohm resistor
T1	500- to 8-ohm audio output transformer

24

*Priority Telephone**

The circuit of this project takes care of two well-known problems occurring when telephones are paralleled in the usual manner. The problems are these:

1. A telephone set can be "bugged" or "jammed" from another set on the same line.
2. The bell in one set may ring when another set is used for dialing out.

The cure for the first problem is easy. One terminal of each telephone set is provided with an exclusive switch built from two four-layer diodes. The recommended 4EX 581 diodes are commerical versions; the 4E 30-8 are low-cost trigger diodes with looser specs.

For a 48-volt system, the diode's breakdown voltage must be about 33 volts. Stable operation requires a maximum holding current of 15 mA, a value easily met in the circuit. The voltage drop across the diodes when on is 0.8 volts.

When a call is answered, the line voltage drops to less than 8-volts and all sets except the one first lifted are cut off. Changing sets during a conversation is possible, of course, but only one set at a time will be

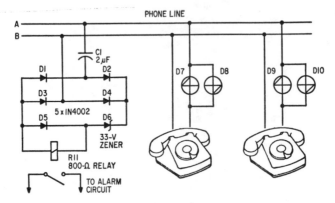

Fig. 24.1   Priority telephone

*From O. Baden, "Give Priority to First Phone Lifted in Parallel-Connected Phone Systems," *Electronic Design* 18, Sept. 1, 1976, p. 110.

"alive." The dial and bell functions of a normal set are not affected by the switch.

The second problem is solved by disconnecting all local bells in the sets and using a common call relay to drive a distributed alarm system. The relay should be time-delayed, or, better still, voltage-level dependent as shown. Incoming ringing signals (80 volts at 20 hertz) are detected immediately, and the commutating noise from dialing is rejected.

The combined circuit (Fig. 24.1) allows paralleling an almost unlimited number of telephones.

Parts List

C1	2-μF capacitor
D1 – D5	1N4002
D6	33-volt zener
D7 – D10	1N3833, EX 581, E 30-8
RY1	800-ohm relay

25

Voice-Activated Telephone Tape Recorder*

Did you ever want to record a telephone conversation and get annoyed at the button pushing required, having other things on your mind? With an automatic recorder you can eliminate this problem. Although these devices are commercially available for about $25, you can build your own for less with the instructions given in the author's previous book, *Telephone Accessories You Can Build* (Hayden Book Co., 1976).

The big disadvantage of automatic recorders is that if the party you're speaking to puts you on hold or goes away from the phone for a few minutes, your automatic recorder wastes tape recording nothing. This problem can be eliminated with the circuit described here. It can be built for less than $10.

The key to the low cost of this project is the use of a commercially available cassette recorder. Virtually all such recorders have a monitor jack. As a rule, the jack is used with an earphone to keep tabs on what is being recorded. However, we can use the fact that an audio signal is present at the jack while in the record mode to eliminate a lot of circuitry otherwise necessary. The best thing of all is that absolutely no modification of the recorder is required. Another big bonus, which your local telephone company will appreciate, is that the circuit does not require a direct connection to the telephone line.

Theory of Operation

Operation of the voice-operated telephone recorder is really very simple. To begin with, a telephone pickup coil is needed. A ring-shaped one that fits over the earphone part of the receiver is recommended because it seems to produce the strongest signal, but almost any telephone pickup coil will be satisfactory. The coil should be connected to the MIC input of your recorder unless the instructions that come with it state differently.

*Adapted from "Circuit Turns On Tape Recorder Only When Sound is Detected," Michael L. Roginsky, *Electronic Design*, Sept. 27, 1975.

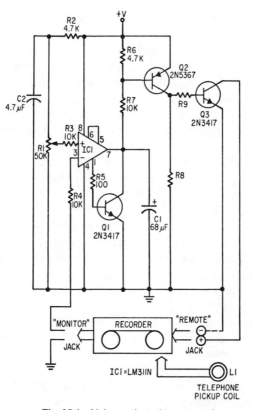

Fig. 25.1 Voice-activated tape recorder

With the tape recorder in the record mode, audio signals are sent to the monitor jack, which is connected to the inverting input of a voltage-comparator integrated circuit (IC1), as shown in Fig. 25.1. The comparator's noninverting input is connected to an adjustable-bias potentiometer (R1), which sets the sound level at which the recorder will start.

At this point, it should be noted that although the recorder is in the record mode, it will not record, the reason being that the motor is turned off by transistor Q3, which controls the remote jack. Some recorders are designed so that both the motor and electronics are shut off when the remote jack is in the off position. If your recorder is of this kind, it will not work with this circuit. To check it out, put an earphone in the monitor jack and put the recorder in the record mode. Place a microphone in the MIC jack and talk into it. You should hear your voice in your earphone. Now turn the recorder off using the remote jack (plug a mini plug into it). Talk into the microphone once again. If you can still hear your voice, your recorder is okay.

When the audio signal from the monitor jack is applied to the comparator, internal circuitry in the IC checks whether the signal is larger than the reference signal (on the noninverting input). If it is not, no signal is allowed to pass to the output. If it is, the amplifier saturates and sends the signal to its output, where an uncommited output transistor is connected in a Darlington configuration with external transistor Q1. From here, the signal is applied to a circuit that permits fast turn-on of the recorder so that little if anything is missed at the beginning. The circuit also allows for slow turn-off so that momentary pauses won't cut it off prematurely. The time delays for the fast turn-on and slow turn-off are determined by R6, R7, and C1.

When no sound is being picked up by the telephone coil, the output of the comparator is low and transistor Q1 is not conducting — a virtually open circuit. This allows capacitor C1 to charge via resistors R6 and R7. When the charging current ceases, transistors Q2 and Q3 also cease to conduct and the remote jack sees an open circuit; thus the motor is turned off.

When sound is picked up by the telephone coil, it is amplified by the amplifier in the tape recorder and fed to the inverting input of the comparator. If the sound level is greater than the threshold set by R1, transistor Q1 is turned on (becoming a virtual short circuit) and rapidly discharges C1, causing transistor Q2 to turn on as well. Q2 inverts and sharpens the signal produced by Q1, thereby turning on Q3, which then shorts the remote jack and turns on the recorder motor. As long as sound above the set threshold is being picked up, C1 stays shorted and the recorder stays on, recording every word that is said. The minute the sound stops, so does the recorder, thus saving tape. It restarts when the conversation does

Building and Testing

Because the circuit is so simple, no special care is needed to build it except to avoid overheating the IC leads. The whole circuit easily fits on a small piece of perforated board.

Care should be taken to use an appropriate recorder, as described earlier. To test the unit, place the pickup coil on the phone and turn on the power to the circuit. Place the recorder in the record mode. Now dial any digit but zero in order to get rid of the dial tone. Start talking into the handset, and adjust the potentiometer (R1) so that the recorder turns on when you start talking and turns off when you stop. It may take about five seconds for the recorder to stop after you cease talking. Once the potentiometer is set, you shouldn't have to set it again unless your telephone connection is so poor that it almost prevents you from hearing the party you're talking to.

Parts List

C1	68-μF electrolytic capacitor (25 WVDC)
C2	4.7-μF capacitor
IC1	LM311N comparator
L1	telephone pickup coil (see text).
Q1, Q3	2N3417
Q2	2N5367
R1	50-kilohm potentiometer
R2	4.7-kilohm resistor
R3, R4, R7	10-kilohm resistor
R5	100-ohm resistor
R6	47-kilohm resistor

26

Automatic Telephone Recorder With Beeper

The legality of recording telephone conversations has always been a fuzzy area. Some say that as long as one party to the conversation is aware that it is being recorded, then it is legal. Others believe that both parties must be informed. To eliminate controversy, many people and organizations simply introduce a tone onto the telephone line about every 10 seconds so that anyone using the telephone will know that the conversation is being recorded.

Recording a conversation can be a pain in the neck, what with the need to find and hook up a telephone pickup coil and then set up the recorder. Even then it doesn't work sometimes because the coil wasn't put in the right location for picking up a strong signal.

These problems can be eliminated. The inexpensive circuit here described generates an automatic tone to warn that conversations are being recorded.

Theory of Operation

The circuit (Fig. 26.1) consists basically of two parts, a controller that turns on the recorder and beeper and a tone generator that produces the beep tones. The controller is really a very simple device, consisting of only four components: a diode, a capacitor, a relay, and a resistor. The controller is connected in parallel with the telephone line, with the anode of the diode connected to the green wire (also known as L1). This is the positive wire of the two. The other wire, usually red (and also known as L2), is negative.

As soon as the controller is connected to the line, relay RY1 snaps closed, opening the two sets of normally closed relay contacts. One set of contacts is connected to a cable with a subminiature plug on it and thus connected to the remote input of a cassette recorder. This makes it possible to start and stop the recorder any time that the phone is answered or hung up.

If you would rather control the ac power going to the recorder, the normally closed (NC) contact used for this switching application can be

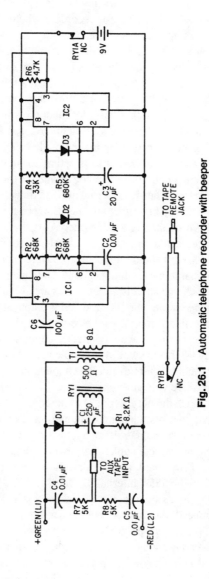

Fig. 26.1 Automatic telephone recorder with beeper

connected in series with the ac power supply to the recorder. This arrangement may be desirable for recorders where the remote switch turns off just the motor and not all the power to the recorder.

The voltage across a telephone line is normally about 48 volts. In the controller circuit, half of that figure is across the 24-volt relay and the other half across the 8.2-kilohm resistor. When a telephone receiver is lifted off the hook, the 48 volts drops down to about 5 or 6 volts. This figure is further divided so that half of it (about 3 volts) is applied to the relay and the other half across the 8.2-kilohm resistor. Since 3 volts is not enough to pull in the relay, or even maintain it in a pulled-in position, the relay releases, and the relay contacts return to their normally closed condition, activating both the tape recorder and the tone beeper.

The tone beeper is fabricated from two 555 timer ICs or from a single 556 dual timer IC. The first timer (IC1) is designed to operate as a 1-kHz oscillator. In this oscillator, or astable, mode of operation, the timer IC, in order to produce oscillation, retriggers itself by connecting the trigger input at pin 2 to the threshold input at pin 6. Diode D2 is used to make the charge and discharge cycles independent, and since R2 and R3 are of equal value, the charge and discharge cycles are also equal, thus producing a square wave signal. The 1-kHz signal is output at pin 3 and coupled to the telephone line through a reverse-connected audio output transformer. If the power supply were connected continuously to pins 4 and 8 on IC1, a steady 1-kHz tone would be coupled into the telephone line, and conversation would be quite difficult. Although pin 8 is connected constantly to the positive voltage supply, pin 4, which is the reset pin, is not. Pin 4 is connected to the output (pin 3) of timer IC2. Pin 4 is used to reset the flip-flop that controls the state of output pin 3 (see Appendix). The reset action occurs when the voltage applied to pin 4 lies between 0 and 0.4 volt. When the voltage applied to pin 4 is greater than 0.4 volt, the reset function is inactivated, and the state of the output pin is determined by other inputs to the 555s, internal flip-flop.

This being the case, it is easy to see why IC1 does not produce a steady 1-kHz tone all the time. When the handset is lifted off the cradle, the voltage on the line drops to about 5 or 6 volts, causing relay RY1 to drop out and the relay contacts to return to their normally closed position. Thus, contact RY1A is closed, applying power to both timer ICs. When power is applied to IC2, timing capacitor C3 is uncharged and the output is high. The output remains high for a period of time determined by the equation:

$$t_1 = 1.1(R4)(C3)$$

or about .75 second. During this period, the 1-kHz oscillator tone is applied to the line. The first pulse is longer than all succeeding pulses. At the end of this period, the output of the IC2 timer goes low and remains that way for a period of time determined by the equation:

$$t_2 = 0.693(R5)(C3)$$

or about 9.5 seconds. During this period, no voltage is applied to pin 4 of IC1, and thus the oscillator does not work and no tone is produced. After the period elapses, once again capacitor C3 of IC2 charges up, and a voltage is applied to pin 4 of IC1, causing a beep tone. The time period of this tone, and all subsequent tones, is determined by the equation:

$$t_3 = 0.693(R4)(C3)$$

or about 0.5 second.

The two oscillators will continue to interact with each other in this manner, producing a half-second tone of 1 kHz every 9.5 seconds until the phone is hung up. When it is, the voltage across the line increases to about 48 volts, relay RY1 is activated, and the normally closed contacts are opened, thereby shutting off both the tape recorder and the beep tone oscillators.

Construction and Checkout

Because of the low frequencies involved, most of the wiring in this project is noncritical. The only thing requiring attention is the connection to the telephone line. Be sure that the wire marked green in the schematic is connected to the green wire (or, failing that, to the wire that is positive) of your telephone line.

Although two integrated circuits are used, a printed circuit board is not vital, and wiring can be done by hand with a piece of breadboard. For convenience, and to facilitate troubleshooting and the changing of parts, it is advised that sockets or Molex pins by used instead of doing the wiring directly to the ICs. Diode D1 can be any diode that has a PIV of 100 volts and can pass 100 mA. The 100-volt PIV is to accommodate the 90-volt ring signal without destroying the diode. Diodes D2 and D3 can be any small-signal silicon diode. The relay can be any low-current, high-resistance relay. The one chosen here is a 24-volt relay with a coil resistance of about 8 kilohms. These figures, however, are not critical. Relays with coil resistances ranging from 2000 to 10,000 ohms can be used so long as R1 is adjusted appropriately. The latter should have a value about equal to that of the coil resistance. Most surplus electronics parts dealers have relays fitting into this category.

For ease in interfacing, it is suggested that either a jack or a modular connector be used to connect this project to the telephone line. But before doing so, it is advisable to check with your local phone company to see if they have any rules forbidding this. Because of the large size of capacitor C1, this project is easily detectable by phone company equipment; phone companies often check for additional equipment by measuring the capacitance across the line. As it stands now, the controller circuit will cause about a 2-second delay before it shuts the tape off, permitting some blank space between conversations. If this space is not important to you, the value of C1 may be considerably reduced.

To check out the device, just connect it to the telephone line and lift the receiver. The first thing you should hear as you connect it to the line is a relay click. Then, when you lift the receiver off the hook, you should hear another relay click. If you do, everything is working fine, and you can proceed to connect the tape recorder to the circuit. Place a tape in the recorder and put it in the record mode. Now lift the receiver off the hook again. The tape should start recording the dial tone. When the phone is hung up, it should stop.

If the circuit does not work as described, you've probably reversed the connections to the green and red wires. Try reversing them again and see since there's nothing you can damage.

Parts List

B1	9-volt battery
C1	250-μF electrolytic capacitor (50 WVDC)
C2, C4, C5	0.01-μF capacitors
C3	20-μF electrolytic capacitor (25 WVDC)
C6	100-μF electrolytic capacitor (25 WVDC)
D1–D3	1N914
IC1, IC2	555 timers
R1	8.2-kilohm resistor
R2, R3	68-kilohm resistors
R4	33-kilohm resistor
R5	680-kilohm resistor
R6	4.7-kilohm resistor
R7, R8	5-kilohm resistor
RY1	24-volt relay (Sigma 65 FIA 824DC)
T1	500- to 8-ohm audio output transformer

27

Phone Sentry Answering Machine

Now you'll have no more missed phone calls with this home-built "Phone Sentry" on the job. It automatically answers the phone for you and takes the caller's message.

The Phone Sentry features an electronic control unit that you can build yourself (see Fig. 27.1). It is used along with one or two unmodified cassette tape recorders for other purposes.

Several rent developments have made the Phone Sentry both possible and practical: inexpensive cassette recorders (endless loop cassettes on which you record your "answer messages"), two new integrated circuits, and highly reliable yet low-cost reed relays. The Phone Sentry makes the equivalent of an expensive phone-answering service available to the smallest businesses and even for home use at a fraction of the cost of units now on the market. If you build the unit yourself, the parts will cost you about $30.

The heart of the Phone Sentry is the electronic control unit that "answers" the phone, sequences the tape recorders, and "hangs up" after the caller's message is recorded. The unit is triggered by the telephone ring. A relay is then energized that turns on the answer tape player. At the end of the answer message, a prerecorded tone signal causes the answer player to be turned off and the message recorder to be turned on. The message recorder remains on for 30 seconds to record the caller's message; then it, along with the entire unit, is automatically turned off, leaving the Phone Sentry ready to receive its next call.

Building the Control Box

Mount the four reed relays, the two integrated circuits, transistor Q1, the neon lamp, capacitors, and resistors on the printed circuit board as shown in Fig. 27.2. Observe capacitor and diode polarities and proper orientation of the IC1 and IC2. Pin 1 of IC1 and IC2 can be determined by the dot or depression in the top of the case. Mount the neon lamp on the foil side of the board, and then bend the leads so that they lie flat against the board.

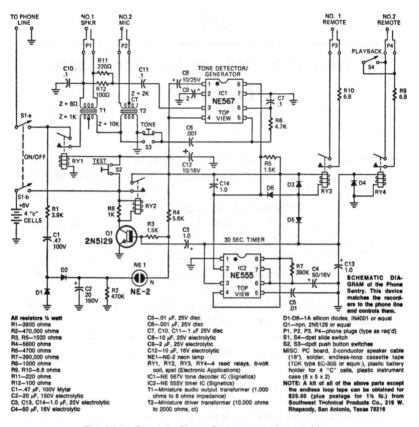

Fig. 27.1   Circuit for Phone Sentry answering machine

Connect 2-foot two-wire leads to T1 and T2 (these will go to plugs P1 and P2) and to RY3; also connect a 6-inch twin lead to RY4 (these will go to P3 and the playback switch). Using three 6-inch leads, connect an spdt pushbutton switch (this will be the tone button) so that the common terminal goes to ground and the normally open terminal to the primary of T2. Connect an spst button switch, using two 6-inch leads, to the test terminal (near R1) and to the normally open contact of S1-6 (on/off switch); this will be the test button.

Connect a 6-foot twisted-pair cable to the common terminal (near C2) and to on/off switch S1-a. Connect the other side of S1-a to the RY1 terminal with a 6-inch lead. Use another 6-inch lead to connect S1-6 to RY2 and then connect the other side of S1-6 to the positive lead from the battery holder. The negative lead from the battery holder connects to common (near C2).

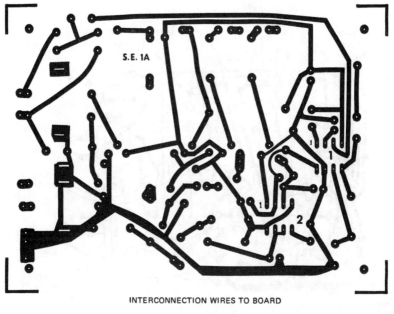

INTERCONNECTION WIRES TO BOARD

Fig. 27.2   Full-size PC board foil pattern

Mount the PC board (Fig. 27.3) and battery holder in the case. Connect a 2-foot twin lead to the N.O. terminals of the playback switch (which goes to P4). The two-wire lead from RY4 is already on these terminals. Bring out the five leads and solder the proper plugs (P1–P4) to their respective cables. Place a piece of white tape around the cables to P1 and P3, and label the tape, ''Answer Unit No. 1.'' Put similar tape on the cables to P2 and P4, and label the tape ''Message Recorder Unit No. 2.'' A four-pin telephone plug, or any plug to match you phone jack, can be mounted on the line cable. Mount the four switches to the front cover of the case.

After you assemble the control unit, record your answer message on the endless loop cassette. The following message is just one example: ''Hello, this is Roger Smith. Your call is being answered by an automatic Phone Sentry because I am either away or unable to answer the phone at this time. However, I would like to return your call. Will you please leave your phone number or message at the sound of the tone, and I will call you back as soon as possible. Remember, leave your phone message at the sound of the tone. Thank you for calling.''

The answer message should be about five seconds shorter than the playing time of the endless loop (30 seconds in this case). Now, insert

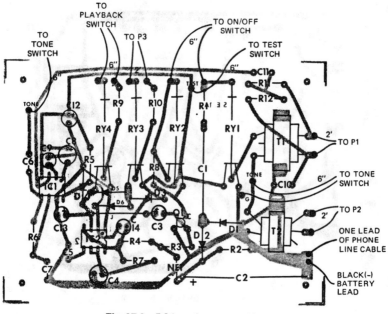

Fig. 27.3 PC board component layout

plug P2 from the Phone Sentry into the microphone jack of answer unit No.1 and plug P3 into the remote jack. Turn the Phone Sentry on, place the recorder in the record mode, and momentarily depress both the test and tone pushbuttons; then turn the Phone Sentry off. Now turn off the recorder and remove P2. You will now have a short tone signal of the proper frequency recorded at the end of your answer message, but be sure that the Phone Sentry is disconnected from the phone line. You may have to check your watch closely when making the recording so that the end of the message or tone doesn't overlap with the beginning.

You are now ready to test your Phone Sentry. Connect the answer player unit No.1 to plugs P1 and P3 (plug P1 to the earphone jack and P3 to the remote jack). Turn on the player to the play mode. Connect message recorder unit No. 2 to P2 and P4 (plug P2 to the microphone jack and P4 to the remote jack). Turn on the recorder to the record mode. With the on/off switch of the Phone Sentry on, depress the test button. The answer player should run for about thirty seconds until the tone signal turns it off and turns the message recorder unit on. The message recorder should also run for about 30 seconds, and then both units should go off. (See Fig. 27.4.)

If all seems well, you can connect the line cable to the red and green (L1 and L2 or T and R) terminals of the phone line. Although your

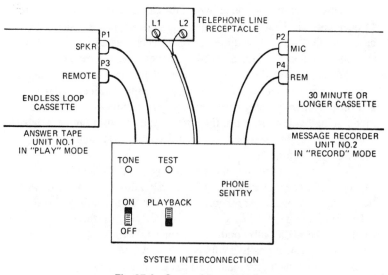

Fig. 27.4 System interconnection

phone junction box may have three or more wires, connect only the two mentioned. Have a friend call you to test the unit. As soon as the Phone Sentry acquires the line, the answer player should start, and you can begin monitoring the call by picking up your receiver. A little experimentation may be necessary to arrive at the proper volume control setting for your cassette recorders.

If your unit doesn't operate properly, check all connections and verify proper plug placement. Be sure that all batteries are new since a balky recorder will cause a malfunction. Check the circuit theory section and schematic to help you track down your trouble. Also, try reversing the leads to the phone line.

Battery life using regular "C" cells should be two months (based on 20 calls per day). You should check your batteries or replace them periodically to insure trouble-free operation. The Phone Sentry can also be used with recorders that have ac battery eliminators.

The endless TDK loop cassette tape is highly recommended for use with the Phone Sentry because it simplifies the operation. Although it costs about $4, the investment is well worth it. Also available are endless loop cassettes of 1 minute, 3 minutes, 6 minutes, and 12 minutes. These might come in handy for special cases where only an announcement is desired. As you may have noticed, any type of tape recorder that has a remote on/off feature can be used as message recorder unit No. 2.

There are cases, of course, where the caller's message does not need to be recorded. Theater, advertising, or store hours announcements are cases in point. The only requirement is to record the announcement with the tone signal at the end; no need to connect message recorder No. 2.

You are now ready to put your Phone Sentry into service. Whenever you leave your business or home, switch the control unit on and place the answer player in the play mode and the message recorder in the record mode. When you return, turn the control unit off, turn on the playback switch, and rewind the message recorder tape. You then can play the messages and return the calls.

The incoming ring signal is rectified by diodes D1 and D2, charging capacitor C2. The present value of C2 ($20 \mu F$) allows the phone to ring for two to three seconds. For longer ring periods, increase the value of C2. When this charge reaches 65 volts — after two or three seconds of ringing — the neon bulb conducts, turning Q1 on. RY2 pulls in and is latched on by R4. A positive voltage of 6 volts is now supplied to the rest of the circuit, causing RY1 and RY3 to pull in. As a result, transformers T1 and T2 are connected to the line and answer player unit No. 1 starts. The output of the answer player is coupled to the telephone lines through T1 and to the input of tone detector 1C1 through capacitor C11.

1C1 contains a phase-locked loop designed for frequency-sensing. It has a controlled oscillator, a phase detector, and a power output stage. When a signal of the same phase as the oscillator is applied to input pin 3, output pin 8 will go low.

When the output of 1C1 goes low, the 30-second monostable 1C2 is triggered and its output (pin 3) goes high, thereby turning on relay 4 and starting the message recorder unit. After the tone, 1C1 goes high and RY3 turns off, stopping the answer player. R7 and C4 were chosen to give unit No. 2 a 30-second on-time. This time can be changed by changing R7 or C4. The caller's message is coupled to message recorder unit No. 2 through T2. At the end of 30 seconds, output pin 3 of 1C2 goes low; this negative-going signal is coupled through C3 to Q1, causing Q1 to turn off and release RY2, thereby de-energizing the circuit.

The Phone Sentry control unit has been designed to provide proper coupling to the phone line when there is no dc voltage on it and the recorders are isolated. The tone signal is below the 2400 Hz required by the phone company. Four "C" cells power the unit.

Parts List

C1	47-μF, 100-V Mylar capactior
C2	20-μF, 150-V electrolytic capacitor
C3, C13, C14	1.0-μF, 25-V electrolytic capacitor

C4	50-μF, 16-V electrolytic capacitor
C5	.01-μF, 25-V disc capacitor
C6	.001-μF, 25-V disc capacitor
C7, C10, C11	.1-μF, 25-V disc capacitor
C8	10-μF, 25-V electrolytic capacitor
C9	2-μF, 25-V electrolytic capacitor
C12	10-μF, 16-V electrolytic capacitor
D1–D6	A-1 silicon diodes (1N4001 or equal)
IC1	NE 567-V tone-decoder IC
IC2	NE 555-V timer IC
NE1, NE2	neon lamps
P1, P2, P3, P4	phone plugs (type as req'd)
Q1	npn transistor (2N5129 or equal)
R1	3900-ohms, ¼-W resistor
R2	470,000 ohms, ¼-W resistor
R3, R5	1500-ohms, ¼-W resistor
R4	5600-ohms, ¼-W resistor
R6	4700-ohms, ¼-W resistor
R7	390,000-ohms, ¼-W resistor
R8	1000-ohms, ¼-W resistor
R9, R10	6.8-ohms, ¼-W resistor
R11	220-ohms, ¼-W resistor
R12	100-ohms, ¼-W resistor
RY1, RY2, RY3, RY4	4 reed relays, 6-volt coil, spst (Electronic Applications)
S1, S4	dpst slide switch
S2, S3	dpdt pushbutton switches
T1	Miniature audio output transformer (1,000- to 8-ohm impedance)
T2	Miniature driver transformer (10,000 to 2000 ohms, ct).

MISC: PC board, two-conductor speaker cable (18 ft), solder, endless-loop cassette tape (TDK type EC–30S or equiv.), plastic battery holder for four "C" cells, plastic instrument case (6 × 5 × 2 in.)

APPENDIX

Getting to Know the 555 Timer*

Integrated circuit (IC) timers are probably the most versatile ICs available today. They can provide precise timing intervals ranging from microseconds to hours. They can also be used as easily controllable, inexpensive oscillators. Applications range from interface circuitry to control systems and alarms.

The most popular of the IC timers available is the 555 timer. It generally comes in an eight-pin minidip package (Fig. A1-A) but is also available in an eight-pin round TO-99 can (Fig. A1-B). In addition, a 14-pin dual in-line package (Fig. A1-C) containing two 555 IC timer chips is available. This is generally designated as the 556 timer, although some manufacturers do use other number designations (XR-2556 and D555) with different pinout configurations (Fig. A1-D).

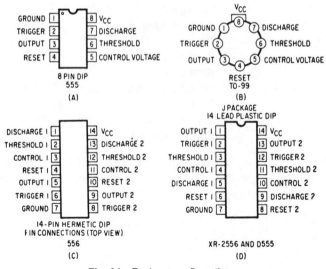

Fig. A1   Package configurations

*From Jules H. Gilder, *110 IC Timer Projects,* Hayden Book Co., Inc.

124

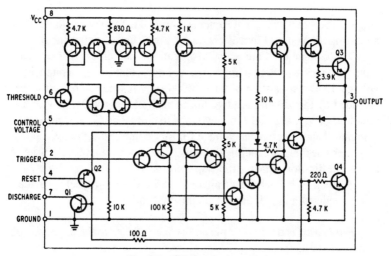

Fig. A2 555 timer schematic

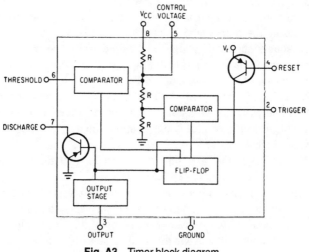

Fig. A3 Timer block diagram

The 555 timer is basically a very stable IC that is capable of being operated either as an accurate bistable, monostable, or astable multivibrator. These three basic modes of operation make the 555 one of the most useful ICs since the operational amplifier.

The 555 IC timer is composed of 25 transistors, 2 diodes, and 16 resistors (Fig. A2) and might appear to be quite complex, but, as can be seen from the block diagram in Fig. A3, the device is functionally quite simple. The circuitry in the IC timer is arranged to form two comparators, a flip-flop, two control transistors, and a high-current output stage.

The comparators are actually operational amplifiers that compare input voltages to internal reference voltages that are generated by a voltage divider consisting of three 5,000-ohm resistors. The references provided by this divider are two-thirds of the supply voltage (V_{cc}) and one-third of V_{cc}. When the input voltage to either one of the comparators is higher than the reference voltage for that comparator, the operational amplifier goes into saturation and thus produces a signal that is used to trigger the flip-flop. And the flip-flop controls the output state of the timer.

A quick look at what each pin on the 555 IC does will help provide a clearer understanding of the 110 circuits that follow.

Pin 1. This is the *ground* pin and gets connected to the negative side of the voltage supply.

Pin 2. This is the *trigger* input. When a negative-going pulse causes the voltage at this point to drop below one-third of the V_{cc}, the comparator to which this input is connected causes the flip-flop to change state, causing the output level to switch from low to high. The trigger pulse must be of shorter duration than the time interval determined by the external R and C. If this pin is held low longer than that, the output will remain high until the trigger input is driven high again.

Pin 3. This is the *output* pin. It is capable of sinking or sourcing a load that requires up to 200 mA of current and can drive TTL circuits. The output voltage available at this pin is approximately equal to the V_{CC} applied to pin 8 minus 1.7V.

Pin 4. This is the *reset* pin. It is used to reset the flip-flop that controls the state of output pin 3. The pin is activated when a voltage level anywhere between 0 and 0.4 V is applied to the pin. The reset pin will force the output to go low no matter what state the other inputs to the flip-flop are in. To prevent unwanted resetting of the output, pin 4 should be connected along with pin 8 to the positive side of V_{cc} when not in use.

Pin 5. This is the *control voltage* input. By applying a voltage to this pin, it is possible to vary the timing of the device independently of the RC network. The control voltage may be varied from 45 to 90 percent of the V_{cc} in the monostable mode, making it possible to control the width of the output pulse independently of RC. When it is used in the astable mode, the control voltage can be varied from 1.7 V to the full V_{cc}. Varying the voltage in the astable mode will produce a frequency-modulated (FM) output. This pin is connected to the internal voltage divider so that a voltage measurement between it and ground should read two-thirds of the voltage applied to pin 8. If, as in most applications, this pin is not used, it should be bypassed to ground to maintain immunity from noise.

Pin 6. This is the *threshold* input. It resets the flip-flop and consequently drives the output low if the voltage applied to it rises above two-thirds of the value of the voltage applied to pin 8. In addition to the voltage level, a current of at least 0.1 μA must be supplied to this pin. This threshold current determines the maximum value of resistance that can be connected between the positive side of the supply and this pin. For 15-V operation, the maximum value of resistance is 20 megohms (MΩ).

Pin 7. This is the *discharge* pin. It is connected to the collector of an npn transistor. The emitter of the transistor is connected to ground, so that when the transistor is turned "on," pin 7 is effectively shorted to ground. Usually the timing capacitor is connected between pin 7 and ground and is discharged when the transistor is turned "on."

Pin 8. This is the *power supply* pin and is connected to the positive side of the supply. The voltage applied to this pin may vary from 4.5 to 16 V for commercial devices. Selected devices that operate at voltages as high as 18 V are available.

The pin numbers and functions for the 8-pin minidip package and the round TO-99 can are identical. The equivalent pin numbers for the 14-pin DIP dual timer are shown in Table A1. Electrical characteristics for the 555 timer are shown in Table A2, while typical operating curves are shown in Fig. A4.

Table A1 Pinout Comparison Chart

Function	555 Timer Pin No.	556 Timer No. 1 Pin No.	556 Timer No. 2 Pin No.	D555 and XR-2556 Timer No. 1 Pin No.	D555 and XR-2556 Timer No. 2 Pin No.
Ground	1	7		7	
Trigger	2	6	8	2	12
Output	3	5	9	1	13
Reset	4	4	10	6	8
Control voltage	5	3	11	4	10
Threshold	6	2	12	3	11
Discharge	7	1	13	5	9
V_{cc}	8	14		14	

Table A2 Electrical Characteristics: $T_A=25°C$, $V_{CC}=+5$ to $+15$ V unless otherwise specified

Parameter	Test conditions	SE 555 Min	SE 555 Typ	SE 555 Max	NE 555 Min	NE 555 Typ	NE 555 Max	Units
Supply voltage		4.5		18	4.5		16	V
Supply current	$V_{CC}=5V$ $R_L=\infty$		3	5		3	6	mA
	$V_{CC}=15V$ $R_L=\infty$		10	12		10	15	mA
Timing error	Low state; Note 1 R_A, R_B=1kΩ to 100 kΩ C=0.1 µF; Note 2							
Initial accuracy			0.5	2		1		%
Drift with temperature			30	100		50		ppm/°C
Drift with supply voltage			0.05	0.2		0.1		%/Volt
Threshold voltage			⅔			⅔		$\times V_{CC}$
Trigger voltage	$V_{CC}=15V$	4.8	5	5.2		5		V
	$V_{CC}=5V$	1.45	1.67	1.9		1.67		V
Trigger current			0.5			0.5		µA
Reset voltage		0.4	0.7	1.0	0.4	0.7	1.0	V
Reset current			0.1			0.1		µA
Threshold current			0.1	.25		0.1	.25	µA
Control voltage level	Note 3							
	$V_{CC}=15V$	9.6	10	10.4	9.0	10	11	V
	$V_{CC}=5V$	2.9	3.33	3.8	2.6	3.33	4	V
Output voltage drop (low)	$V_{CC}=15V$							
	$I_{sink}=10$ mA		0.1	0.15		0.1	.25	V
	$I_{sink}=50$ mA		0.4	0.5		0.4	.75	V
	$I_{sink}=100$ mA		2.0	2.2		2.0	2.5	V
	$I_{sink}=200$ mA		2.5			2.5		V
	$V_{CC}=5V$							
	$I_{sink}=8$ mA		0.1	0.25				V
	$I_{sink}=5$ mA					0.25	.35	V
Output voltage drop (high)	$I_{source}=200$ mA $V_{CC}=15V$		12.5			12.5		V
	$I_{source}=100$ mA $V_{CC}=15V$	13.0	13.3		12.75	13.3		V
	$V_{CC}=5V$	3.0	3.3		2.75	3.3		V
Rise time of output			100			100		nsec
Fall time of output			100			100		nsec

Notes: [1] Supply current when output high typically 1 mA less.   [2] Tested at $V_{CC}=5$ V and $V_{CC}=15$ V.
[3] This will determine the maximum value of $R_A + R_B$. For 15-V operation, the max total $R=20$ MΩ.

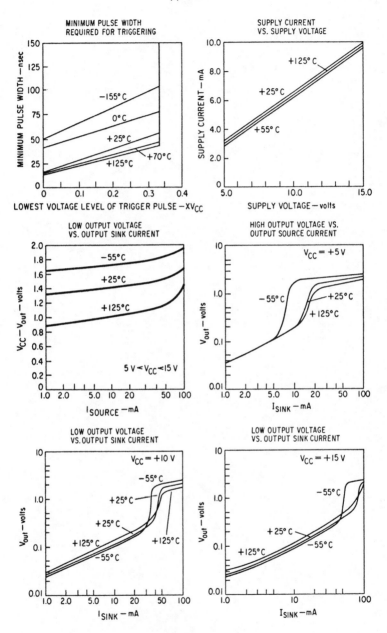

Fig. A4 Typical operating curves